Oxford Lecture Series in
Mathematics and its Applications 9

Series editors
John Ball Dominic Welsh

OXFORD LECTURE SERIES IN MATHEMATICS AND ITS APPLICATIONS

1. J. C. Baez (ed.): *Knots and quantum gravity*
2. I. Fonseca and W. Gangbo: *Degree theory in analysis and applications*
3. P. L. Lions: *Mathematical topics in fluid mechanics, Vol. 1: Incompressible models*
4. J. E. Beasley (ed.): *Advances in linear and integer programming*
5. L. W. Beineke and R. J. Wilson (eds): *Graph connections: Relationships between graph theory and other areas of mathematics*
6. I. Anderson: *Combinatorial designs and tournaments*
7. G. David and S. W. Semmes: *Fractured fractals and broken dreams*
8. Oliver Pretzel: *Codes and algebraic curves*
9. M. Karpinski and W. Rytter: *Fast parallel algorithms for graph matching problems*
10. P. L. Lions: *Mathematical topics in fluid mechanics, Vol. 2: Compressible models*

Fast Parallel Algorithms for Graph Matching Problems

Marek Karpinski

Department of Computer Science
University of Bonn

and

Wojciech Rytter

Department of Computer Science, University of Warsaw
and
Department of Computer Science, University of Liverpool

CLARENDON PRESS • OXFORD
1998

Oxford University Press, Great Clarendon Street, Oxford OX2 6DP
Oxford New York
Athens Auckland Bangkok Bogota Bombay
Buenos Aires Calcutta Cape Town Dar es Salaam
Delhi Florence Hong Kong Istanbul Karachi
Kuala Lumpur Madras Madrid Melbourne
Mexico City Nairobi Paris Singapore
Taipei Tokyo Toronto Warsaw
and associated companies in
Berlin Ibadan

Oxford is a trade mark of Oxford University Press

Published in the United States
by Oxford University Press, Inc., New York

A catalogue record for this book is available from the British Library

Library of Congress Cataloging in Publication Data
(Data available)

ISBN 0 19 850162 5

Typeset by the authors

Printed in Great Britain by
Bookcraft Ltd., Midsomer Norton, Avon

Preface

This book is a comprehensive and a selfcontained introduction to the basic methods of designing efficient parallel algorithms for graph matching problems.

Recently, a growing interest has been observed in the areas of randomized, algebraic and geometric techniques in the design of efficient combinatorial algorithms. The available textbooks in these areas, however, are rather rare, and treat them either generally or in a rather restricted way as small parts of more general texts on algorithms and data structures. In this book, we intend to approach the area in a novel way, and provide a straightforward algorithmic introduction together with a set of example problems specially designed for the student.

Our main goal is to bring together combinatorial, algebraic and probabilistic approaches and show how this works in the development of efficient parallel algorithms for the interesting combinatorial example of the *matching problem*. This is, for several reasons, probably the best problem to serve as a common platform for demonstrating the usefulness of combining the above approaches.

Our second goal is to provide a simple representation of many interesting algorithms by means of examples and illustrations. We hope that the text will be easily understandable to a student.

The matching problem is one of the central problems in graph theory as well as in the theory of algorithms; there is even an advanced monograph devoted exclusively to this combinatorial problem, see [LP86]. The problem is closely related to other graph problems, for example *maximal independent sets*, *edge-colorings*, *f-matchings*, *network flows*, *etc.* Therefore we will also present parallel techniques for these and other problems.

The problem is also particularly interesting in the algorithmic sense, as there are many polynomial time (and highly nontrivial) sequential algorithms. That the matching problem is in the class $\mathcal{P}$ (class of problems solvable in polynomial time) is not obvious. It is included in $\mathcal{P}$ thanks to the ingenious introduction of nontrivial combinatorial tools such as *alternating paths* and *blossoms*. The matching problem itself is somewhere on the boundaries of the class $\mathcal{P}$. There are versions which are computationally inherently difficult, see for example [KW88]. The computation of the number of all maximum matchings is complete in the class called $\#\mathcal{P}$, which is at least as hard as an *NP*-class of problems. Some of its decision variants become *NP*-complete problems. Hence it is no surprise that the parallelization of sequential algorithms for the matching problem is difficult, yet both algorithmically and combinatorially very interesting.

The problem is especially interesting in the parallel setting. It is not known if it has a deterministic *NC*-solution, where *NC* is the class of so-called *well-*

parallelizable problems (meaning problems solvable in polylogarithmic parallel time with polynomially many processors). It is generally believed that this problem is in *NC*, and there is even an (algebraic) algorithm which is a good candidate (presented in this book)!

The matching problem becomes parallelizable (in the randomized sense) using probabilistic and algebraic tools (determinants and special polynomials). There are several *randomized NC* (so called *RNC*) algorithms. Though the problem has a strictly combinatorial flavor, its parallel algorithmics are inherently related to algebra and geometry. It is quite interesting how classical mathematics has become closely related to the new areas motivated by the modern technologies. On the other hand these new areas have stimulated and pushed some subjects of classical mathematics in new directions. A good example is the emergence of sparse algebra and geometry in the late 1980s which was directly motivated by some parallel algorithmic matching applications. There are no known *RNC*-algorithms for the general matching problem which avoid the algebraic determinantal approach.

The main theoretical tools of this text are combined into three independent chapters (combinatorial tools, probabilistic tools, and algebraic tools), and the subsequent text concerns only the structure and analysis of the algorithms presented.

As well as the large diversity of problems related to matchings, there are also many interesting subclasses of graphs for which parallel matching is easier. The techniques used for the matching problem are much more general. This combination of many approaches makes the matching problem an attractive and fascinating subject and a meeting point for many interesting algorithmic techniques as well as new algebraic and geometric areas.

We would like to thank many colleagues for their contributions to the present book at various stages, especially Piotr Berman, Norbert Blum, Arthur Czumaj, Elias Dahlhaus, Krzysztof Diks, Andrzej Dorf, Carsten Dorgerloh, Anna Gambin, Dariusz Kowalski, Larry Larmore, Thomas Lickteig, Mike Luby, Wojtek Plandowski, Alistair Sinclair, Roman Smolensky, Umesh Vazirani, Dominic Welsh, Avi Wigderson, Jürgen Wirtgen and Alex Zelikovsky. We also thank Angus Macintyre for his continuing encourangement during the process of writing. We have worked on this book at varying times and places, at the University of Bonn, the International Computer Science Institute, Berkeley, the University of Warsaw, Princeton University, and the Université de Paris Sud. We owe a great deal to the students and teaching assistants of the respective courses we taught at the Universities of Bonn and Warsaw using earlier versions of the manuscript. Their detailed feedback has lead to several improvements and simplifications in the final text.

The work on this book was partially supported by the DFG Grant Bo 56/142-1, KBN Grant 8T11C01208, by the ESPRIT BR Grants 7097 (RAND), and EC-US 030 (RAND–REC), by the Volkswagen–Stiftung, DIMACS, and the Max–Planck Research Prize.

We are deeply indebted to Christine Marikar for her continuous help and support throughout the whole writing-effort. Special thanks go to Harald Deuer, Thomas Lang, Martin Lies and Andrea Lohmüller for technical support in producing the final LaTeX file, and several LaTeX macro packages along the way.

It was a pleasure to work with Soenke Adlung from Oxford University Press on this project.

Bonn *M.K.*
Liverpool *W.R.*
November 1997

Contents

1

Introduction

1.1 Matchings in graphs: terminology, examples

A *graph* is a useful abstraction of many real world situations. It occurs when a problem is essentially related to the structure of interconnections between some objects. For example, a road map can be viewed as a graph: the interconnected objects are cities and connections correspond to roads. The connections can be *one-way* or *two-way*, and the corresponding graph can be directed or undirected.

Another real world example is a group of people, the connections between them corresponding to the relation "the person x knows the person y". In abstract graph terminology the interconnected objects (cities, persons) are called *vertices* or *nodes* (these terms are equivalent terms in this book). The connections are called *edges*. In the case of relation "x knows y" we have an edge (x, y).

A graph G is formally given by a pair $G = (V, E)$, where V is the set of vertices and E is the set of edges. The edges can be unordered pairs of different vertices: the graph is called *undirected*; or ordered pairs: the graph is *directed*. We consider only the case when all edges are undirected or all edges are directed. Also we do not allow multiple edges between the same vertices (in the directed case between the same ordered pair). We assume that V is finite and its cardinality is usually denoted by n. This is the main parameter related to the size of the graph. The second parameter is the number m of edges of a given graph.

We assume untill the end of this section that considered graphs are undirected. If $e = (x, y)$ is an edge then we say that x and y are its endpoints. We say that two edges (x, y) and (x', y') are disjoint iff they have no common endpoint; in other words $\{x, y\} \cap \{x', y'\} = \emptyset$.

A path is a sequence of distinct nodes of V; each two consecutive nodes are joined by an edge. A cycle is a similar sequence of nodes, the only difference is that the first and the last nodes are equal.

Any set of pairwise disjoint edges is called a *matching* of a given graph. Denote by $\text{MatchNum}(G)$ the size (number of edges) of a maximum cardinality matching of a graph $G = (V, E)$ (cf. Figure 1.1). This number is called the *matching number*. A *perfect matching* is a matching of size $|V|/2$.

The basic problems in the algorithmics of matchings can be stated as follows:

existence: for a given graph G compute $\text{MatchNum}(G)$.
construction: compute any matching of maximum cardinality.

Obviously the computation of all matchings of maximum cardinality is *compu-*

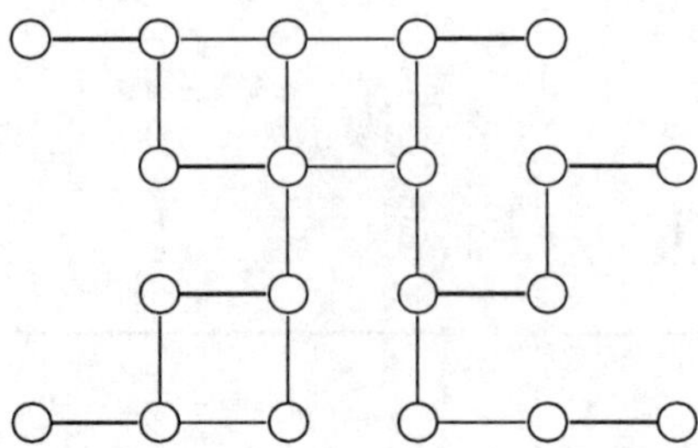

Fig. 1.1 An undirected graph G and a maximum cardinality matching (bold lines). MatchNum(G) = 8. Why is there no larger matching? How many matchings of cardinality 8 are possible?

tationally infeasible since there can be exponentially many such matchings, for example if G is a *complete graph* or if it is the graph whose structure is shown in Figure 1.2.

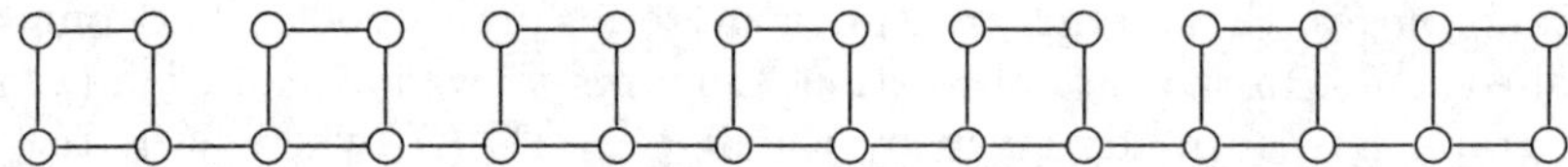

Fig. 1.2 This graph has exactly 2^7 perfect matchings.

Although we cannot list exponentially many matchings in polynomial time, their number can be written with polynomially many digits. However, we do not know of any general polynomial time algorithm that can compute such a number and we have no reason to believe that such an algorithm exists.

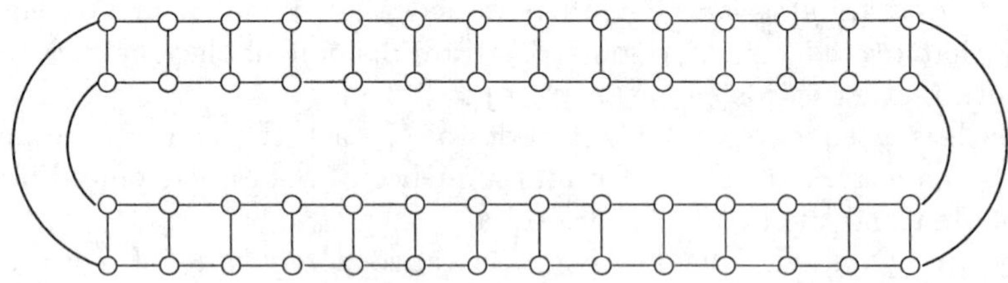

Fig. 1.3 The problem of counting the number of all perfect matchings is rather difficult; no polynomial time algorithm is known to solve it and probably none exists. Calculate this number for the graph above.

The matching problem appears quite naturally in the case of *bipartite graphs*. The bipartite graph is formally given by a triple $G = (X, Y, E)$, where X, Y are *disjoint* sets of vertices, and there are only edges between vertices in X and vertices in Y. Usually $|X| = |Y|$. An example of bipartite graph is illustrated in

Figure 1.4.

One real life situation related to the bipartite matching problem is the following: assume we have n workers and n machines and a relation which says which worker can operate on which machine. We ask if there is an assignment of workers to machines such that each machine is serviced. In other words we ask for a perfect matching in a bipartite graph $G = (X, Y, E)$, where X is the set of workers and Y is the set of machines. The problem is to find a good matching between jobs and workers.

There is another formulation of the bipartite matching known as a *marriage problem*. Assume we have a set X of boys and a set Y of girls, and a symmetric relation E "who knows whom". We want to create the maximum number of married pairs consistent with the relation E (of course the real life problem is much more complicated).

The bipartite matching problem can also be formulated in terms of the *systems of distinct representatives* (*sdrs*, in short). Assume that we have a family $X = (x_1, x_2, \ldots, x_n)$ of finite sets. We wish to find a system $\mathcal{S}$ of pairs $\{(x'_i, y'_i) : y'_i \in x'_i\}$, where all x'_i are all given sets, each one appearing at most once, and all y'_i are pairwise different. y_i is called the representative of x_i with respect to $\mathcal{S}$. The system $\mathcal{S}$ is called a system of distinct representatives. The question about the maximum cardinality *sdr* is equivalent to the question about the maximum cardinality matching in the corresponding bipartite graph, see the example in Figure 1.4.

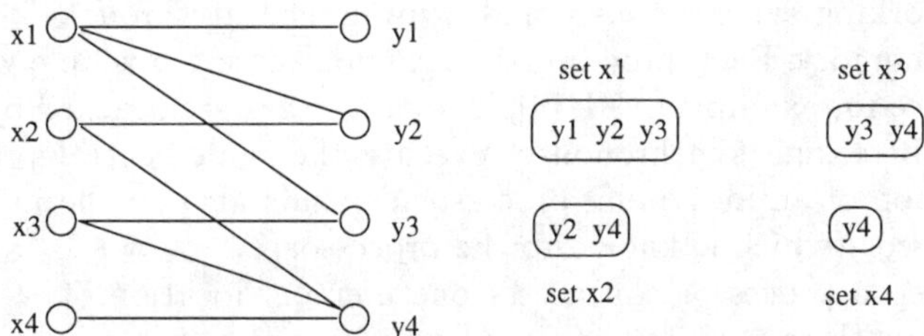

Fig. 1.4 A family of sets and a corresponding bipartite graph. The matching $\{(x1, y2), (x3, y4)\}$ corresponds to a system of representatives, in which $y2$ is the representative of $x1$ and $y4$ is the representative of $x3$. Observe that this matching is inclusion maximal.

Bipartite matching is closely related to maximal flows in networks. In the case of sequential computations the nonbipartite matching is much more complicated than the bipartite case. It can no longer be viewed as the problem about systems of representatives and does not correspond directly to the maximum flow problem.

Nonbipartite matching appears for example in the situation when we have n persons and a symmetric relation between them and we have to join them in pairs for some reason. There are also many other applications.

We only show here how nonbipartite matching applies to the problem of scheduling two processors. Assume we have two processors, each capable of performing one job in one time unit. Suppose we have n jobs and a partial ordering (given by an acyclic directed graph G), which describes the *precedence constraints* for the jobs. If there is a directed edge from x to y then y can be performed only after x is completed (not necessarily immediately after that). Let G^* be the *independency* graph of G defined as follows: there is an edge $(x, y) \in G^*$ iff there is no directed path from x to y in G. It can be shown that the minimal time to complete the given set of jobs equals $n - \text{MatchNum}(G^*)$, see [FKN69] and [CG72].

1.2 The model of parallel computations: PRAM

A very general model of parallel computations will be used in this book. The first reason for using such a model is that we are interested mostly in exposing the parallel nature of some problems related to matchings in graphs, without going into technical details related to the real machine hardware (like the synchronization and communication problems) as this is not the aim of this book. The second reason is that the main problem related to matchings concerns the existence of so-called NC-algorithms, and the class of NC-algorithms does not depend on any particular model, like for example the *hypercube* network.

The *parallel random access machine* (*PRAM*, in short) became a standard model for the *presentation* of parallel algorithms. PRAM consists of a number of processors working synchronously and communicating through a common random access memory. Each processor is a random access machine with standard operations, see for example [AHU74]. The processors are indexed by consecutive natural numbers, and synchronously execute the same central program. However, the action of an individual processor depends also on the index (number) of this processor (which is known to the processor).

In one step a processor can access one memory location. The basic PRAM model differs with respect to simultaneous access of the same memory location by more than one processor. We allow many processors to read simultaneously from the same location, but we do not allow many different processors to write simultaneously to the same memory location. Such a model of PRAM is known as a CREW PRAM. The abbreviation CREW stands for "Concurrent Reads Exclusive Writes", since concurrent reads are possible and concurrent writes are excluded. This is the submodel of PRAM used in this book.

The parallel algorithms are presented in a similar way as in *Efficient parallel algorithms* [GR88], and *An introduction to parallel algorithms* [J92]. There is no accepted universal language for the presentation of parallel algorithms. On the other hand it would be rather difficult to present these algorithms adequately using languages oriented towards any existing hardware of parallel computers. The PRAM model is widely accepted in the literature on parallel algorithms.

The parallelism is presented in this book by the following type of informal statement:

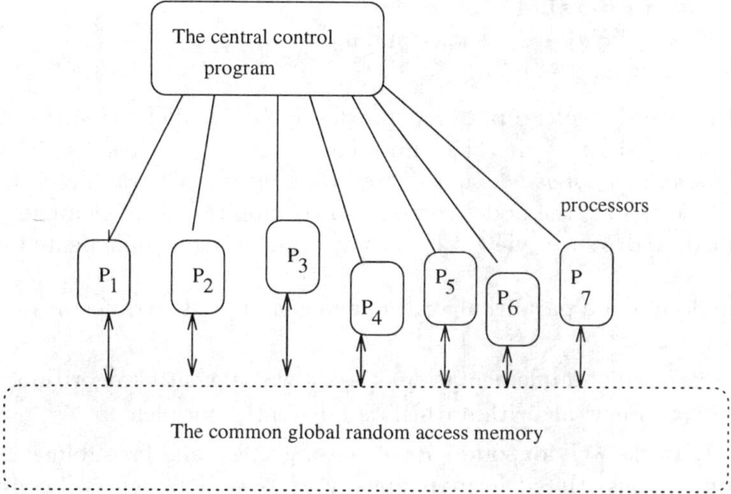

Fig. 1.5 A schematic illustration of the parallel random access machine

for all $x \in X$ **do in parallel** action(x).

Execution of this statement consists of:

(1) assigning a processor to each element of X;

(2) executing simultaneously by all assigned processors the operations specified by action(x).

Usually the part $x \in X$ looks like "$1 \leq x \leq m$" if X is an interval of integers $[1 \ldots m]$. In graph algorithms it can also have the form "for each $v \in V \ldots$, where V is the set of a vertices of a given graph.

The PRAM model is well suited to work with tree-structured objects or tree-like recursive computations. We demonstrate how it works on the following basic problem: compute the sum of n elements $a_1, \ldots, a_n$. Assume for simplicity that n is a power of two.

```
Algorithm Parallel_Sum;
    for k := 0 to log(n) − 1 do
      for each 1 ≤ i ≤ n do in parallel
        if i − 2^k ≥ 1 then a_i := a_i + a_{i−2^k};
      return a_n.
```

At first glance it is hard to see the tree-like structure. However, the algorithm uses a strategy of an *implicit* balanced binary tree. The sums are being processed bottom up, level by level, for each element of one level simultaneously.

The above algorithm can also be described in more informal and recursive way:

compute in parallel

$$a_{n/2} := a_1 + \ldots + a_{n/2} \text{ and } a_n := a_{n/2+1} + \ldots + a_n;$$

return $a_n = a_{n/2} + a_n$.

In other words we compute independently in parallel the sums of the first and the second half of the elements. Then we add the results. This type of parallel *divide and conquer* is quite often used in parallel graph algorithms. The structure is tree-like, the nodes correspond to subintervals of elements which are to be added, and we can view it also as a recursive tree implemented iteratively (level by level).

In the design of a parallel algorithm one can apply two different methodological approaches:

(1) make a parallel implementation of a known sequential algorithm;

(2) construct a new algorithm which is inherently parallel.

Method (1) works well for some subclasses of graphs, and for sublinear time parallel computations related to matchings in graphs. However, NC-computations for the general matching problem use the second method, since known sequential algorithms (based on alternating paths) have a very sequential structure. Some parts can be parallelized, and the time can be improved by using many processors, but this does not give a polylogarithmic time algorithm for the whole problem. New algorithms had to be developed, mostly reducing the graph-theoretic problem to an algebraic problem related to the problem of computing determinants and the rank of matrices.

1.3 The complexity classes NC and RNC

The main aim of parallel computing is the decrease of computation time. Basically we are interested in a polylogarithmic time ($\log^k n$, for a fixed k). However, such an improvement in time complexity is done at the expense of employing many processors. From a practical point of view the number of processors cannot be too large. We assume that the *feasible* number of processors is polynomial in n.

From the point of view of parallel computation the most important class of problems is the class NC of problems computable in polylogarithmic time with polynomially many processors. Denote by $\mathcal{P}$ the class of problems computable in sequential polynomial time. Perhaps the most important question in the theory of parallel computations is the question:

is $\mathcal{P} = NC$?

It is our strong belief that the answer to this question is negative; however, the question could be of similar difficulty to the "$P = NP$" problem. The matching problem is in $\mathcal{P}$ and the main question related to parallel computations and the matching problem can be formulated as:

is the matching problem in NC?

There is a general belief that the answer to this question is affirmative, and the solution would use some advanced algebraic techniques.

There is a large class of problems which are in $\mathcal{P}$ and which are believed to be not in *NC*. These problems form the class of so-called $\mathcal{P}$*-complete* problems. We define the problem here to be $\mathcal{P}$*-complete* if it is in $\mathcal{P}$ and for any other problem in $\mathcal{P}$ there is an *NC*-reduction to this problem. There are many natural graph-theoretic problems which are known to be $\mathcal{P}$-complete, like maximal flows and lexicographically first depth-first search. There are also some variations of the matching problem which are $\mathcal{P}$-complete.

There are also problems in the class *CC* which are between *NC* and $\mathcal{P}$-complete problems. The class *CC* was introduced in [MS92]. One of the *CC*-complete problems is the *greedy matching* problem. The greedy algorithm computes a matching by processing a list of edges sequentially: an edge is included if it has no common endpoint with edges included up to that moment. The produced matching is not necessarily of maximum cardinality. However, its computation is complete in the class *CC*.

Despite the fact that the parallelization of the general matching problem looks very difficult there are many subclasses of graphs for which we have *NC*-algorithms for the matching problem. We will present several interesting algorithms for such subclasses.

A closely related problem to matchings is that of (inclusion) *maximal independent sets*. We show that both problems of inclusion maximal matchings and inclusion maximal independent sets are in the class *NC*, despite the fact that the *greedy* sequential approach is probably not parallelizable.

There is still another important parallel complexity class: *RNC*. This is the class *Randomized NC* of problems possessing randomized polylogarithmic time algorithms with polynomially many processors. One of the main results related to matchings and shown in this book is:

the maximum matching problem is in RNC.

1.4 Bibliographic notes

The basic material on matchings can be found in Lovász and Plummer [LP86] and in any standard text on graphs, see Berge[B73], and on graph algorithms, for example Gibbons [Gi85]. The books of Aho, Hopcroft and Ullman, see [AHU74], and of Kozen, [K92], are nice references for the complexity and analysis of algorithms. For the models of parallel computation and basic techniques we refer to Gibbons and Rytter [GR88].

2
Combinatorial tools

Recall that MatchNum(G) is the size (number of edges) of a maximum cardinality matching of a graph $G = (V, E)$ and a perfect matching is a matching of size $|V|/2$. There are two basic combinatorial tools needed to compute this number:

augmenting paths (to construct a matching of a maximal size k);

witness sets (to provide an NC-proof that there is no matching of size larger than k).

Almost all efficient sequential algorithms use *augmenting paths*, contrary to NC-algorithms. On the other hand *witness sets* are not used frequently in sequential computations but they are needed in parallel randomized computations. If an algorithm returns a matching of cardinality k then we are not certain if this cardinality is the maximum possible, even if there is a very small probability of failure. Our algorithm is a *Monte Carlo* algorithm.

A better situation occurs when the algorithm provides a maximum cardinality matching and an easy (checkable in NC) proof that it is maximum. In this case the algorithm becomes the so-called *Las Vegas* algorithm. If a matching is *perfect* then it gives the optimality proof by itself.

2.1 Augmenting paths

Assume $M \subseteq E$ is a (not necessarily maximum) matching in a given graph $G = (V, E)$. The edges of M are called *matched* and other edges are called *free*. Similarly, nodes which are endpoints of edges in M are called *matched* (in M), and all other vertices are called *free* (in M).

An *augmenting path* w.r.t. M is a path in G such that its edges are alternately *matched* and *free*, and endpoints of the path are *free*, see Figure 2.1.

If π is an augmenting path w.r.t. M then certainly M is not a maximum matching. We introduce the operation Augment(M, π) which returns a matching of cardinality larger by one. The operation works by exchanging free and matched edges on the path π. Hence new matching is larger and M is not a maximum matching.

It was shown by Berge, see [B73], that the converse implication is valid:

> if M is not a maximum matching then there exists an augmenting path.

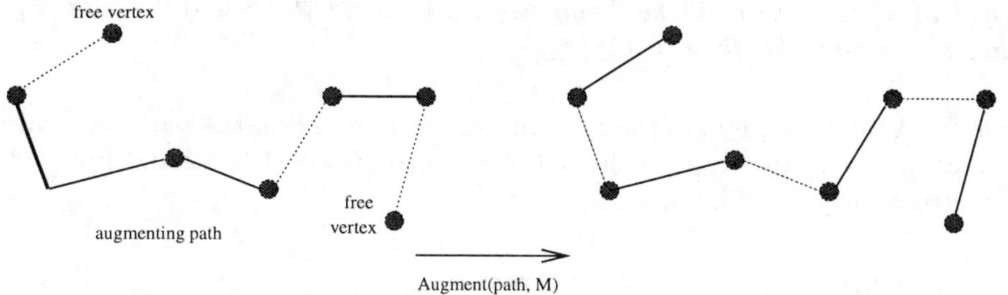

Fig. 2.1 Example of an augmenting path and the matching after augmenting. Free edges correspond to dotted lines.

This fact is fundamental for the construction of sequential polynomial time algorithms.

Remark How to employ this fact in *NC* computations of maximum matchings is a challenge.

The Berge theorem is the first point of the theorem below; it was strengthened (as the second point of the theorem below) by Hopcroft and Karp [HK73].

Theorem 2.1.1
(1) *A matching M is a maximum cardinality matching iff there is no augmenting path with respect to M.*
(2) *Assume* M, M^* *are matchings and* $k = |M^*| - |M| > 0$. *Then there is a set of k vertex disjoint augmenting paths w.r.t. M and the shortest path has length at most* $|V|/k - 1$.

Proof Consider the graph $G' = M \oplus M^*$, where $\oplus$ is the *symmetric difference* of sets. G' consists of a collection of several disjoint cycles and paths. The difference in cardinality between M and M^* on each cycle is zero, and on each path is $+1$, 0 or -1, so there should be at least k paths where this difference is in favor of M. All these paths are augmenting paths and disjoint. It follows, by a simple counting argument, that the shortest of them cannot have more than $|V|/k$ vertices, so cannot have length (number of edges) larger than $|V|/k - 1$.

□

As a direct consequence of point (2) of Theorem 2.1.1 we have the following useful result.

Lemma 2.1.2
Let M be a matching, and M^* *be a maximum cardinality matching of G. If the length of the shortest augmenting path w.r.t. M is at least* $\sqrt{n}$ *then* $|M^*| - |M| \leq \sqrt{n}$.

Lemma 2.1.3
Assume π_1 *is a shortest augmenting path w.r.t. a matching* M_1 *and* π_2 *is any*

augmenting path w.r.t. the matching $M_2 = \text{Augment}(M_1, \pi_1)$. *If the paths* π_1, π_2 *have a common edge then* $|\pi_2| \geq |\pi_1| + 1$.

Proof Assume $u1$, $u2$ are the first and the last vertex on the path π_1 common to both paths π_1, π_2. Assume the paths π_1, π_2 are from $v1$ to $v2$ and from $w1$ to $w2$, respectively, see Figure 2.2.

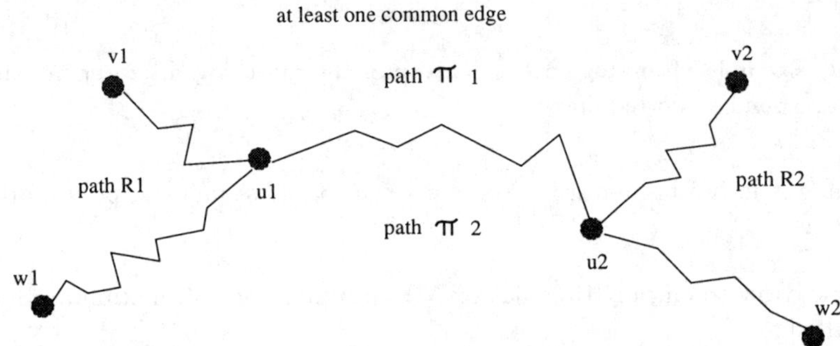

Fig. 2.2 Two overlapping paths π_1, π_2 imply two other paths $R1, R2$.

Then there are two additional paths $R1, R2$ which go through the vertices $(v1 \ldots u1 \ldots w1)$ and $(v2 \ldots u2 \ldots w2)$, respectively. It can be proved that $R1, R2$ are augmenting paths w.r.t. M; we omit an easy proof. Assume the length of π_1 is k. Then the lengths of each of these additional paths are also at least k, since π_1 was a shortest path. Now it is easy to see that π_2 has to have at least one more edge than π_1, since π_1 and π_2 have a common edge, and the contribution of paths $R1, R2$ to π_2 is at least k.

□

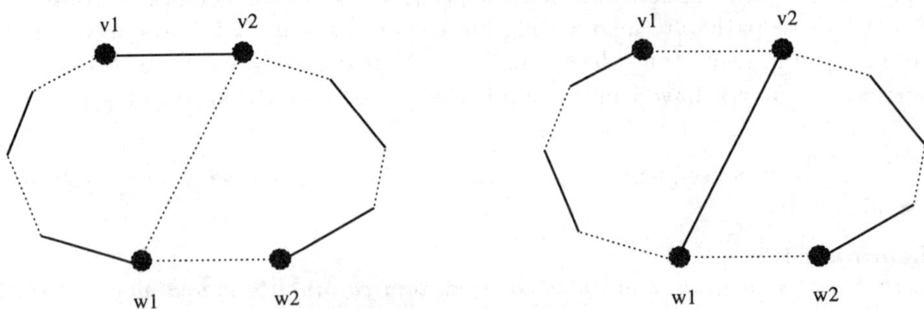

Fig. 2.3 The left picture shows an alternating cycle C in which free edges are dotted satisfying condition $\Psi_C(v1, v2, w1, w2)$. The matching is changed on this cycle as indicated in the right picture.

The lemma implies the following theorem.

Theorem 2.1.4
Let $\mathcal{F}$ be a (inclusion) maximal family of disjoint shortest augmenting paths with respect to a matching M. Then the length of a shortest augmenting path with respect to $M \oplus \mathcal{F}$ increases.

Assume that C is an alternating cycle w.r.t. M, and consider the situation shown in Figure 2.3. Define the following condition $\Psi_C(v1, v2, w1, w2)$:

$v1, v2, w1, w2$ are on C in the order $v1, v2, w2, w1$,
$(v1, v2) \in M$, $(w1, w2) \notin M$, and there is an edge $(w1, v2)$ in G.

Lemma 2.1.5
If $\Psi_C(v1, v2, w1, w2)$ then we can change the matching on the cycle C by adding $(w1, v2)$ to the matching in such a way that the cardinality of the matching does not change.

Proof The suitable modification of the matching M is demonstrated in Figure 2.3.

□

We mention briefly here a *weighted version* of the matching problem. The idea of augmenting paths can also be used successfully in this case. Assume each edge e has a real value weight(e) associated with this edge. A matching is a maximum weight matching (of a given cardinality k) iff its total weight is maximum among all possible matchings of G of cardinality k. Observe that to deal with minimum weight matchings it is enough to change the signs of each initial weight. If π is an augmenting path (in the unweighted sense) then define the *incremental value* $\Delta(\pi, M)$ of π as the difference between the weights of free and matched edges w.r.t. M on π. After applying the operation Augment(M, π) we have a new matching of total weight weight$(M) + \Delta(\pi, M)$. We can extend Theorem 2.1.1 to the weighted case as follows.

Theorem 2.1.6
Assume M is a maximum weight matching of cardinality k and π is an augmenting path which maximizes $\Delta(\pi, M)$. Then Augment (M, π) is a maximum weight matching of cardinality $k + 1$.

Proof We refer to [K92].

□

2.2 Deficiencies and witness sets

The *witness sets* are related to a global deficiency and a local deficiency. Let G be an undirected graph. Its (global) *deficiency* (DEFECT(G), in short) is the (minimum) number of *unmatched* vertices in a maximum matching of G:

$$\text{DEFECT}(G) = |V| - 2 \cdot \text{MatchNum}(G)$$

Observation MatchNum$(G) = n/2 -$ DEFECT$(G)/2$ and the computation of the *matching number* of a graph is equivalent to the computation of its deficiency. We shall see later that the computation of the deficiency can be reduced to the computation of so-called *witness sets*.

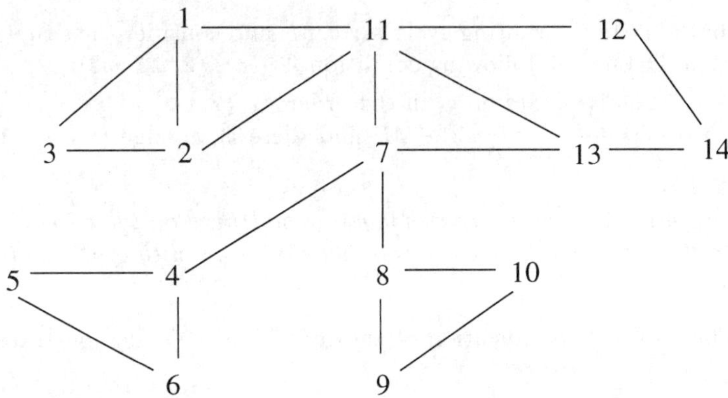

Fig. 2.4 The set $X = \{7, 11\}$ provides a short proof that this graph has no perfect matching:
$|X| = 2$ and after removing X we shall have four *odd components*. DEFECT$(G) = 4 - 2$ and MaxNumMatch$(G) = (|V| -$ DEFECT$(G))/2 = (14 - 2)/2 = 6$.

For a graph G' denote by $\theta(G')$ the number of *odd components* (connected components of odd cardinality) of G'. If X is a set of vertices of G then let $G - X$ be the graph which results after removing all vertices X together with edges incident to them.

Denote

$$\text{LOCAL\_DEFECT}_G(X) = \theta(G - X) - |X|.$$

The number LOCAL_DEFECT$_G(X)$ is called here the *deficiency of the set* X in G. Observe that the deficiency of a set can be negative. In particular if $X = \emptyset$ then the condition LOCAL_DEFECT$_G(X) > 0$ means that $|V|$ is odd, and obviously a graph with an odd number of vertices cannot have a perfect matching. Assume we want to convince somebody quickly that a given graph G has no perfect matching. It is enough to present a set (*witness* against perfect matching) X which satisfies

$$\text{LOCAL\_DEFECT}_G(X) > 0.$$

Assume $|X| = k$ and $\theta(G - X) = j$. Then $j > k$ and G has no perfect matching. This follows from the fact that there are more than k vertices (in odd components of $G - X$) which can only have a partner in X.

The impossibility of a perfect matching follows from the fact that there are not sufficiently many potential partners in X. In fact this argument proves the following stronger property.

Lemma 2.2.1
DEFECT$(G) \geq$ LOCAL_DEFECT$_G(X)$.

The set X is called a *witness set* of G iff DEFECT$(G) =$ LOCAL_DEFECT$_G(X)$.

Example It is very easy to convince somebody that the graph G presented in Figure 2.4 has no perfect matching. Take $X = \{7, 11\}$. Then $|X| = 2$ and $\theta(G - X) = 4$. Hence G cannot have a perfect matching.

Observation Assume MatchNum$(G) = k$. If X is a witness set for G then X gives an *easy NC*-proof that the graph cannot have a larger matching than k, due to Lemma 2.2.1 and the fact that the value $\theta(G - X)$, and consequently the number LOCAL_DEFECT$_G(X)$, is easily computable in parallel by (well-parallelizable) *connected component* algorithms.

Remark The existence of witness sets is nontrivial and even surprising.

2.3 Bipartite graphs and their witness sets

The matching problem for bipartite graphs is also called a *marriage problem* since the following formulation is possible: we have a set A of boys and a set B of girls and a symmetric relation between them of *acquaintance*. Assume the acquaintance relation is described by the edges of a corresponding bipartite graph $G = (A, B, E)$. We want to set the largest number of married couples with the restriction that both partners in each couple know each other (are related through an edge of our graph).

In the case of a bipartite graph $G = (A, B, E)$ we introduce the *left deficiency*, denoted LEFT_DEFECT(G) (deficiency with respect to the part A), of G as the number of *unmatched* vertices in A in a maximum matching of G.

Hence LEFT_DEFECT(G) is the minimum number of married boys (in A) in a matching. If $|A| > |B|$ then obviously LEFT_DEFECT$(G) > 0$.

If we have a set X of boys who together know less than $|X|$ girls then obviously there is no matching in which all boys are married. If there is no such matching then there is a corresponding set X. This fact is called the *marriage theorem* and this is one of the facts in a series of *min–max* theorems in combinatorics.

Denote by Neigh$_G(X)$ the set of neighbors of X in a given graph G; usually we omit the subscript $_G$. Let $G = (A, B, E)$ be a bipartite graph and $X \subseteq A$. Then

$$\text{LOCAL\_DEFECT}_G(\text{Neigh}_G(X)) = |X| - |\text{Neigh}_G(X)|.$$

Observation Obviously LEFT_DEFECT$(G) \geq$ DEFECT$_G(X)$ for each $X \subseteq A$.

If the bipartite graph has a perfect matching then $|A| = |B|$. Bipartite graphs with $|A| = |B|$ are called *equibipartite graphs*.

We say that an equibipartite graph $G = (A, B, E)$ is *equisaturated* iff it has no perfect matching, but adding any new edge between A and B (between existing vertices) implies the existence of a perfect matching.

All equisaturated graphs have a very special structure presented in Figure 2.5. Denote by $Q(A1, A2, B1, B2)$ the bipartite graph $G = (A, B, E)$ such that A, B are disjoint unions $A1 \cup A2$ and $B1 \cup B2$, and there are all possible connections between the vertices from the pairs of sets $(A1, B1)$, $(A1, B2)$ and $(A2, B2)$. The structure of such a graph is illustrated in Figure 2.5.

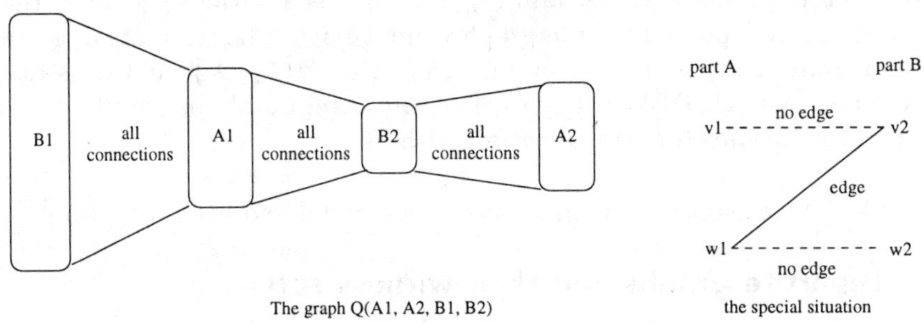

Fig. 2.5

Lemma 2.3.1
If G is an equisaturated graph then $G = Q(A1, A2, B1, B2)$, where $|A2| > |B2|$, for some sets $A1, A2, B1, B2$.

Proof Let us use the *marriage* formulation, where A are boys and B are girls. By assumption, $|A| = |B|$. There should be a boy who does not know everybody because otherwise G would possess a perfect matching. Let $A2$ be the set of all boys with this property.

Claim If a girl knows one boy from $A2$ then she knows all boys from $A2$.

Assume the claim is false. Then we should have the *special situation* shown in Figure 2.5. G is equisaturated, and hence adding an edge $(v1, v2)$ implies the existence of a perfect matching M containing $(v1, v2)$.

On the other hand adding $(w1, w2)$ implies some other perfect matching M' containing $(w1, w2)$ and not containing $(v1, v2)$. Take the symmetric difference $M \oplus M'$. Since both matchings are perfect it implies an alternating cycle C w.r.t. M.

If C does not contain $(w1, w2)$ then we can exchange on C the edges of M with free edges and obtain a perfect matching for the original graph. We have a contradiction, since G is equisaturated. Otherwise we have the situation which appears in Lemma 2.1.5. M does not contain the edge $(w1, w2)$. We change the matching according to Lemma 2.1.5 and we obtain a perfect matching for G. Again we have a contradiction. This proves the claim.

Denote by $B2$ the set of all girls such that each boy from $A2$ knows exactly the girls from $B2$. Denote also $B1 = B - B2$ and $A1 = A - A2$.

Obviously $|A2| > |B2|$, since otherwise we would have a perfect matching in G.

□

There are several different proofs for the marriage theorem; our proof is *structural* and prepares the reader for the same type of structural proof of Tutte's theorem in the next section.

Theorem 2.3.2 (marriage theorem)
Assume $G = (A, B, E)$ is an equibipartite graph. Then
(1) *G has no perfect matching iff* $\text{LOCAL\_DEFECT}_G(X) > 0$ *for some* $X \subseteq A$.
(2) $\text{LEFT\_DEFECT}(G) = \max\{|\text{LOCAL\_DEFECT}_G(X)| : X \subseteq A\}$.

Proof Assume G has has no perfect matching. We add the largest possible number of edges between A and B in such a way that G still has no perfect matching. Then the first point follows directly from Lemma 2.3.1, since now the graph should have the structure $Q(A1, A2, B1, B2)$ described in the lemma. Then we can take $X = A2$ and we have $|\text{Neigh}_G(X)| < |X|$.

The second point of the theorem follows by adding $\text{LEFT\_DEFECT}(G) - 1$ new vertices to B and connecting them to all vertices in A, and the same number of new vertices to A and connecting them to all of B. Then we apply the first point to the newly created graph.

□

We can also formulate the theorem in terms of witness sets.

Theorem 2.3.3 (witness theorem for bipartite graphs)
Let $G = (A, B, E)$ be a bipartite graph. Then there is an A-witness set $X \subseteq A$.

Define the *vertex cover* as a set V_c of vertices such that each edge of G is incident to a vertex in V_c. This concept is *dual* to maximum matching in bipartite graphs in the following sense.

Theorem 2.3.4 (bipartite duality)
Assume G is bipartite. Then the maximum size of a matching in G equals the minimum size of a vertex cover.

Proof Let X be an *A-witness* set of $G = (A, B, E)$ and let Y be the set of neighbors of X. Take $(A - X) \cup Y$ as a required minimum vertex cover.

□

2.4 Tutte's theorem and witness sets in general graphs

If a graph G has a set X such that $\text{LOCAL\_DEFECT}(X) > 0$ then certainly G has no perfect matching. Surprisingly, if there is no set X which implies (in the sense above) that G has no perfect matching then G has a perfect matching. This beautiful result (for general graphs) was first shown by Tutte [T47]. We sketch

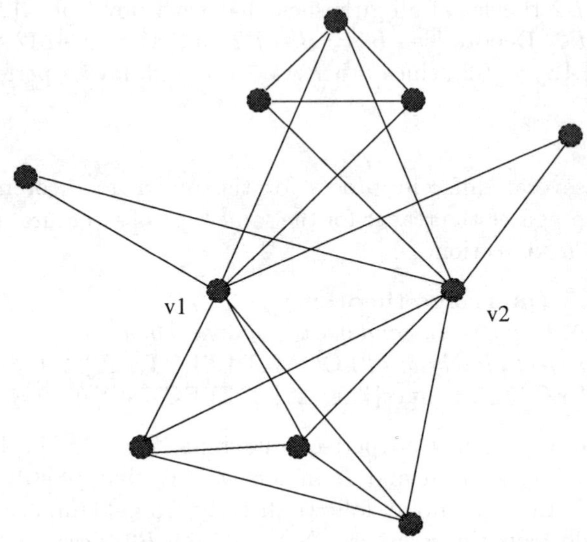

Fig. 2.6 The graph STAR (2,1,1,3,3). The *center* of such a "generalized star" is $\{v1, v2\}$.

a shorter proof given by Lovász, see [LP86]. We start with a *special structure lemma.*

Denote by K_m a complete graph with m nodes. Let STAR $(i, p_1, p_2, \ldots, p_j)$ be a graph which consists of K_i with each vertex connected (by an edge) to each other vertex of disjoint copies of $K_{p_1}, K_{p_2}, \ldots, K_{p_j}$, and there are no edges between these disjoint copies.

The graphs of this type are called here *generalized stars.* The clique K_i is called a *center* of the generalized star. An example of STAR (2,1,1,3,3) is illustrated in Figure 2.6.

A graph with an even number of vertices and without perfect matching is called *saturated* iff the addition of any new edge implies the existence of a perfect matching. All saturated graphs are very similar and they have a structure like the graph in Figure 2.6, owing to the following lemma.

Lemma 2.4.1 (special structure lemma)
Assume G is a saturated graph. Then G is a generalized star. There is a number k and $k+2$ odd numbers $p_1, \ldots, p_{k+2}$ such that G is isomorphic to STAR $(k, p_1, p_2, \ldots, p_{k+2})$.

Proof Let V_0 be the set of vertices of degree $|V|-1$; denote $k = |V_0|$. We prove that V_0 is the center of a *star* satisfying the thesis.

Claim 1 (harder part) All connected components of $G - V_0$ are complete graphs.

Assume there is a component G_i of $G - V_0$ that is not a complete graph. It is easy to see that there are in $G - V_0$ four distinct vertices v_1, v_2, w_1, v_2 which

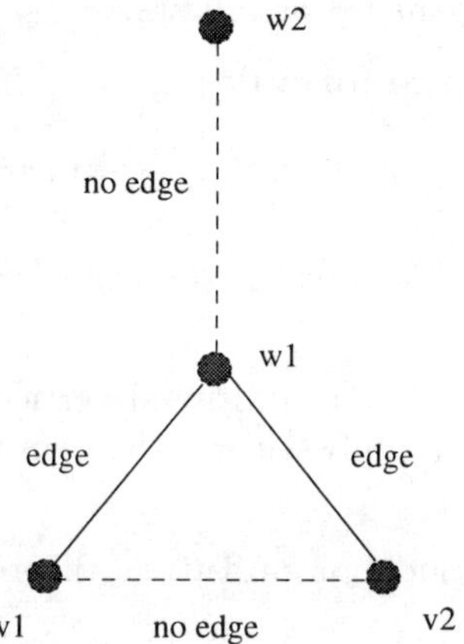

Fig. 2.7 If G is not a generalized star then there is the *special situation* for four vertices.

satisfy the situation shown in Figure 2.7. Similarly as in the proof of Lemma 2.3.1 we can consider two perfect matchings M,M' resulting after the introduction of a new edge $(v1, v2)$ or a new edge $(w1, w2)$, respectively. Again we obtain an alternating cycle C w.r.t. M which satisfies condition $\Psi_C(v1, v2, w1, w2)$, shown in Figure 2.3. This leads to a contradiction (with the nonexistence of perfect matching in G) as in the proof of Lemma 2.3.1. This completes the proof of the claim.

Claim 2 There are $k+2$ components of $G - V_0$ and all of them are *odd.*

Once we know that G is a generalized star then it is easy to see that all components of $G - V_0$ are odd and their number is larger than k. The negation of this subclaim would contradict the fact that G is saturated or that G has no perfect matching. Owing to parity arguments there are at least $k+2$ components in $G - V_0$.

□

Theorem 2.4.2 (Tutte's theorem)
A graph $G = (V, E)$ has a perfect matching iff $\text{LOCAL\_DEFECT}_G(X) \leq 0$ *for each* $X \subseteq V$.

Proof From Lemma 2.2.1 it is enough to show that if G has no perfect matching then there is a set X such that $\text{LOCAL\_DEFECT}_G(X) > 0$. Add to G the largest number of new edges (between the original vertices) such that the graph still has no perfect matching. Then, the new graph G' becomes saturated and it is a generalized star STAR $(k, p_1, p_2, \ldots, p_{k+2})$ according to Lemma 2.4.1. Then we can set X to the center of this star and thus $\text{LOCAL\_DEFECT}_{G'}(X) > 0$, since $|X| = k$ and $\theta(G - X) = k + 2$. It is easy to see that this inequality still holds if we remove the newly created edges.

□

The theorem can be strengthened as follows.

Theorem 2.4.3 (the Berge formula)

$$\text{DEFECT}(G) = \max\{\text{LOCAL\_DEFECT}_G(X) : X \subseteq V\}.$$

Consequently, the set which maximizes the right hand side of the above equality is a witness set.

Proof Add $(n - \text{DEFECT}(G))$ additional vertices, and connect them to all original vertices of G. Then apply Tutte's theorem to the new graph.

□

An example of the application of Tutte's theorem is the following fact, see [LP86] for the proof.

Corollary 2.4.4 (Petersen's theorem)
If a 3-regular graph has at most two bridges then it has a perfect matching.

Open problem Construct an NC-algorithm for perfect matchings in graphs satisfying assumptions of Petersen's theorem. The theorem guarantees that a perfect matching exists but it seems that its proof (via Tutte's theorem) is purely existential.

Up to now our proofs about witness sets have been rather existential. Now we show the construction of a concrete and very special witness set for each graph. We say that a vertex $v \in V$ is *critical* iff v is *matched* in each maximum cardinality matching of G, in other words

$$v \text{ is } \textit{critical} \text{ iff } \text{MatchNum}(G) > \text{MatchNum}(G - x).$$

We define

$$\text{TutteSet}(G) = \{v \in V : v \text{ is } \textit{critical} \text{ and has a } \textit{noncritical} \text{ neighbor in } G\}.$$

The theorem below gives *explicitly* a witness set.

Theorem 2.4.5 (witness theorem)
TutteSet(G) *is a witness set of* G.

Proof The proof is a side effect of the *structural algorithm* presented in the next chapter.

□

Example Let us look at the graph $G1$ in Figure 2.8. Take the set $X = \{15, 16\}$ indicated in the figure. After deleting X we have four odd components. $G1$ has 26 vertices.

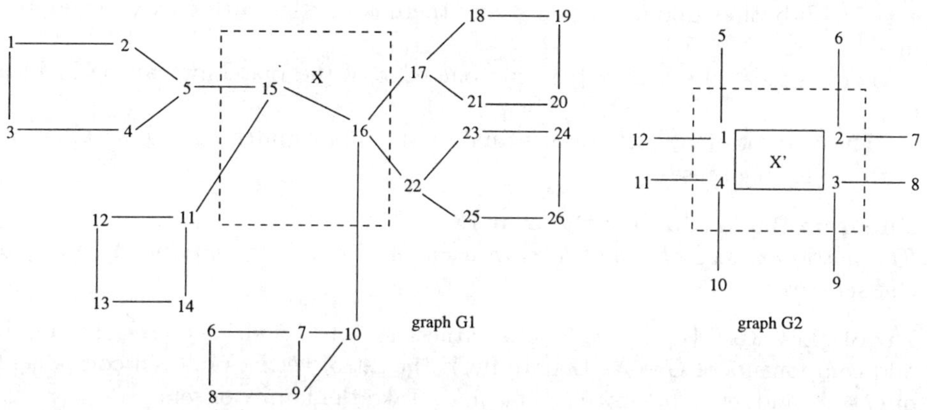

Fig. 2.8 Example graphs $G1$, $G2$. Let $X = \{15, 16\}$, $X' = \{1, 2, 3, 4\}$, then $\text{LOCAL\_DEFECT}_{G1}(X) = \theta(G1 - X) - |X| = 2$, and $\text{LOCAL\_DEFECT}_{G2}(X') = \theta(G2 - X') - |X'| = 4$.

It follows from the relation between deficiency and the matching number that the maximum matching of $G1$ has at most $(26 - (4-2))/2 = 12$ edges. In fact we could find such a matching. Hence X is a Tutte set for $G1$. Similarly X' is a Tutte set for $G2$.

We have

$$\text{DEFECT}(G1) = 2 \text{ and } \text{DEFECT}(G2) = 4.$$

The critical vertices in $G1$ are $\{1, 2, 3, 4, 5, 11, 12, 13, 14, 15, 16\}$. We have that

$$\text{TutteSet}(G1) = \{16\}$$

since in the graph $G1$ 16 is the only critical vertex with a noncritical neighbor; in fact it has three such neighbors. Observe that

$$\text{LOCAL\_DEFECT}_{G1}(\{16\}) = 2.$$

Also observe that $\text{TutteSet}(G1) \neq \{15, 16\}$, so we have two different witness sets for $G1$.

In the graph $G2$ only vertices 1, 2, 3, 4 are critical, each of them having a noncritical neighbor. Hence we have $\text{TutteSet}(G2) = \{1, 2, 3, 4\}$.

2.5 Odd-set covers

Define the *odd-set cover* as any family $\mathcal{F}$ of odd cardinality subsets of V such that each edge of G is incident to a vertex in a one-element set in $\mathcal{F}$ or has both endpoints inside a set belonging to $\mathcal{F}$. The cost of $\mathcal{F}$ is the sum of costs of sets in $\mathcal{F}$, where the cost of a $(2k+1)$-element set is defined to be $\max\{k, 1\}$.

It is easy to see that $\text{MatchNum}(G)$ cannot be larger than the cost of an odd-set cover. Indeed for each $2k+1$ set Z in $\mathcal{F}$ there are at most k disjoint edges with both endpoints inside Z and there is no edge with only one endpoint in Z.

For a singleton set $Z = \{v\}$ only one edge of the matching can be incident to v.

The concept of odd-set cover is dual to maximum matching in general graphs in the following sense.

Theorem 2.5.1 (duality theorem)
The maximum size of a matching in a graph G equals the minimum cost of an odd-set cover.

Proof Let $X = \{v_1, \ldots, v_k\}$ be a witness set of G and let $O_1, O_2, \ldots, O_p$ be odd components of $G - X$. Denote by Y the set of vertices in even components of $G - X$ and let w be any element of Y. Take the family of sets

$$\mathcal{F}_{odd} = (O_1, O_2, \ldots, O_p, \{v_1\}, \{v_2\}, \ldots, \{v_k\}, Y - \{w\}, \{w\})$$

Then it is not difficult to see that $\mathcal{F}_{odd}$ is a required minimum vertex cover, owing to the definition of the witness set.

□

2.6 Good-edges lemma

In this section we introduce the concepts of "good" vertices and "good" edges. The abstract notion of *φ-good* vertices and edges will be useful later when we shall deal with maximal matchings, independent sets and so-called f-matchings. Let $G = (V, E)$ be an undirected graph. Assume $\mathcal{R}$ is any linearly ordered set and we have a function

$$\varphi : V \to \mathcal{R}$$

We say that a vertex v is *φ-good* (see Figure 2.9) iff $\varphi(w) \leq \varphi(v)$ for at least $\lceil \deg(v)/3 \rceil$ neighbors w of v.

An edge is said to be *φ-good* iff at least one of its endpoints is *φ-good*.

Lemma 2.6.1 (good-edges lemma)
At least half of the edges in the graph are φ-good.

Proof Let us direct all edges in the graph. If (v, w) is an undirected edge such that $\varphi(v) > \varphi(w)$ we put the direction from v to w; if $\varphi(v) = \varphi(w)$ then we fix any (unique) direction for (v, w).

The number of incoming (outgoing) edges for the vertex v is denoted by $\text{indeg}(v)$ ($\text{outdeg}(v)$) for each node v.

By the definition, all bad edges end in bad vertices and each bad vertex has its indegree not greater than $\frac{1}{2}$ times its outdegree, so we have

$$m = \sum_{v \in V} \text{outdeg}(v) \geq \sum_{\substack{v \text{ is bad} \\ v \in V}} \geq 2 \cdot \sum_{\substack{v \text{ is bad} \\ v \in V}} \text{indeg}(v) \geq 2 \cdot \text{ number of bad edges.}$$

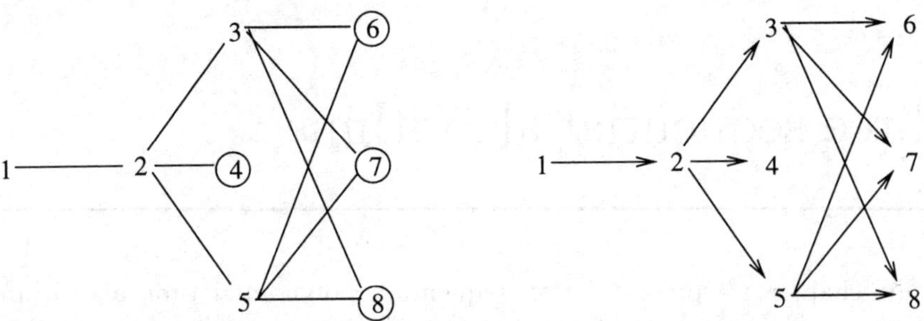

Fig. 2.9 Let $\varphi(i) = i$. Then φ-good nodes are circled in the graph G on the left. All edges are φ-good except (1,2), (2,3) and (2,5). The graph on the right shows G after directing each edge as in the proof of Lemma 2.6.1.

Hence there are at least $m/2$ φ-good edges.

□

2.7 Bibliographic notes

Most of the material on the combinatorics of matchings is contained in the main monograph on this subject: the book by Lovász and Plummer [LP86]. We also refer to Berge [B73].

3

Three sequential algorithms

In this chapter we present basic, sequential, polynomial time algorithms for matchings. The existence of any polynomial time algorithm for matchings in general graphs is a nontrivial fact.

We discuss why these algorithms are hard to parallelize (in the sense of *NC* implementations). The first algorithm (of Hopcroft and Karp, see [HK73]) works efficiently for bipartite graphs and the two others work for general graphs. All of them have many variations which are quite interesting from the point of view of sequential computations. However, these variations would deserve a separate monograph and, as our main concern is parallel complexity, we concentrate only on basic sequential algorithms. They use augmenting paths as a main tool.

A general structure of an algorithm computing a maximum cardinality matching by using the augmenting paths approach is as follows.

General augmenting-paths algorithm;
$M := \emptyset$;
while there is an augmenting path π
  do $M :=$ Augment(M, π);
return M.

In the case of weighted matching we choose in the algorithm a path π of maximum *incremental weight* and this gives a weighted version of the augmenting-paths algorithm which we call the Weighted-Augmenting-Paths algorithm.

In one step the matching is enlarged by one and we can have a linear number of such steps. Observe that such algorithms looks *inherently sequential* and no *NC* algorithm probably can be directly derived from it. However, there is possibly some limited parallelism in this algorithm, since we can still parallelize one stage (computation of an augmenting path). This was done by Schieber and Moran in [SM86] for general graphs and by Goldberg, Plotkin, Vaidya in [GPV93] for bipartite graphs.

We present the sublinear time algorithms of Goldberg–Plotkin–Vaidya and Schieber–Moran later in this book. The main point in the parallel implementation of augmenting-path algorithms is to find in parallel polylogarithmic time one augmenting path or in parallel sublinear time many disjoint paths. In the latter case we can apply many *disjoint* operations Augment in parallel.

3.1 Hopcroft–Karp algorithm

The computation of augmenting paths is much easier if a graph is bipartite, so we start with the bipartite case. Throughout this section we assume G is a bipartite graph $G = (A, B, E)$.

Assume M is a matching of G. Construct a directed graph $\vec{G} = \mathrm{Digraph}(G, M)$ by directing all matched edges from A to B and other edges from B to A. Denote by A_0, B_0 the sets of free vertices in A and B, respectively. The following lemma is obvious.

Lemma 3.1.1
A bipartite graph G has an augmenting path iff there is a directed path in $\vec{G} = \mathrm{Digraph}(G, M)$ from A_0 to B_0.

The lemma directly implies a simple $O(n^3)$ time implementation of the augmenting-paths algorithm for the bipartite case, since we can search for a suitable path by applying (for example) the DepthFirstSearch (DFS, in short). For convenience we can add a *special* vertex v^* which has edges leading to all free vertices in A. An augmenting path is found by a DFS, which starts at this *artificial* vertex v^* and searches for any free vertex in B. It takes $O(n + m)$ time, which is $O(n^2)$. We need to find augmenting paths only $O(n)$ times, and hence all together the complexity is $O(n^3)$.

The $O(n^3)$ time algorithm was improved by Hopcroft and Karp by augmenting in one stage simultaneously *along* all paths in *any* maximal (with respect to inclusion) set $\mathcal{P}$ of disjoint augmenting paths of (the same) shortest length. Denote any such set $\mathcal{P}$ by $\mathrm{MaxSetShortPaths}(A_0, B_0)$.

```
Hopcroft_Karp algorithm;
M := ∅;
repeat
    construct the graph G⃗ = Digraph(G, M);
    {A0, B0 are the sets of free vertices in A, B};
    P := MaxSetShortPaths(A0, B0);
    for each π ∈ P do
        M := Augment(π, M)
until P = ∅ {there is no augmenting path};
return M;
```

Lemma 3.1.2
The Hopcroft–Karp algorithm makes at most $2 \cdot \sqrt{n}$ iterations.

Proof An easy consequence of Lemma 2.1.2 and Theorem 2.1.4 from Chapter 2. Observe that Lemma 2.1.2 was especially formulated and proved only for the proof of this theorem.

□

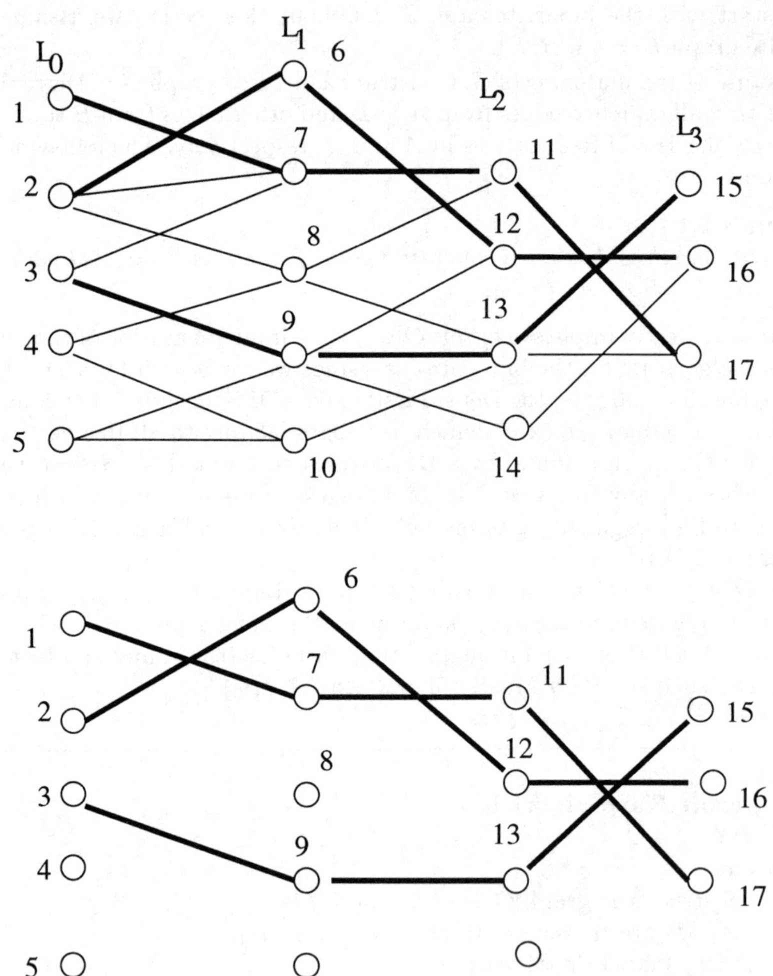

Fig. 3.1 A set of disjoint paths constructed by the function MaxSetShortPaths. Assume Partial_DFS first goes to vertices with smaller labels. $L_0 = A_0$ and $L_4 \subseteq B_0$.

Lemma 3.1.3
Let A_0, B_0 be disjoint sets of vertices in a directed acyclic graph $\vec{G}$. Then we can compute MaxSetShortPaths(A_0, B_0) *in time $O(n^2)$.*

Proof We employ first the BreadthFirstSearch (BFS, in short) to find the length k of a shortest path from A_0 to B_0 and to produce the sequence of disjoint *layers*

$$A_0 = L_0, L_1, L_2, \ldots, L_k \subseteq B_0,$$

where, for $0 \leq i < k$, L_i is the set of vertices at distance i from A_0 and L_k is the set of nodes in B_0 at distance k from A_0.

Then we use the following *partial* procedure DFS(v) which works like classical DFS, but stops in the first moment of finding a vertex in L_k. The maximal set of disjoint paths is found through a repeated use of the function Partial_DFS. The function MaxSetShortPaths (described later) returns the required set of paths. □

```
function Partial_DFS(v);
{v ∈ L0}
start at v and using DFS find some vertex w ∈ Lk;
in the first moment when a vertex in Lk is found
stop and return a path from v to w;
if there is no such path then return a special value nil;
all visited vertices are removed from the graph;
```

Example Figure 3.1 illustrates how the function MaxSetShortPaths works on an example layer graph. We call Partial_DFS(v) with arguments $v = 1, 2, 3$ successively; it returns paths from 1 to 17, from 2 to 16 and from 3 to 15. Then Partial_DFS(4) and Partial_DFS(5) both return the value *nil*. The final set of disjoint paths is shown in the figure.

```
function MaxSetShortPaths(A0, B0);
PathEdges := ∅;
compute the sequence of layers L0, L1, L2, ..., Lk; {use BFS}
for each v ∈ L0 do
  begin
    π := DFS(v);
    if π ≠ nil then PathEdges := PathEdges ∪ {π};
  end
return
    the set of all paths consisting of edges in PathEdges;
```

Lemma 3.1.2 and Lemma 3.1.3 directly imply the following theorem.

Theorem 3.1.4
There is an $O(n^{2.5})$ time sequential algorithm for maximum bipartite matchings.

We will see later that both the BreadthFirstSearch and computation of MaxSet-ShortPaths can be parallelizable to some extent. This leads to a sublinear time parallel implementation of the algorithm of Hopcroft and Karp, which is a topic of Chapter 4.

3.2 Edmonds algorithm and blossoms

In this section we present the Edmonds algorithm [E65]. Its main feature is the introduction of a special auxiliary structure called the *blossom*.

Let G be a general undirected graph. In this general case we cannot apply the construction from the preceding section and use the graph $\vec{G}$, since we do not know now how to direct the edges of G. Nevertheless the structure of the Edmonds algorithm is similar. It finds an augmenting path w.r.t. the current matching M and updates the matching using the same procedure Augment as before.

The basic operation is FindPath(v, M) which finds any augmenting path starting in v. If there is no such path then the result is the empty path.

Assume that Augment$(\emptyset, M) = M$, observe that if v is matched (in M) then

$$FindPath(v, M) = \emptyset.$$

```
Algorithm General_Matching;
M := ∅;
for each v ∈ V do
    π := FindPath(v, M);
        M := Augment(π, M)
return M;
```

There is one *suspicious* point in the algorithm above. Each point is considered as a possible starting point of an augmenting path only *once*. The theorem below states that this is correct.

Theorem 3.2.1
The algorithm General_Matching *is correct.*

Proof We have to prove that once we consider the vertex v as a possible starting point of an augmenting path then we do not need to consider it again in this sense. Of course if v is the starting point of such a path, then v will never be free, and there is no sense in considering it again as a starting point.

We have only to consider the case when we checked v for the first time and found that no augmenting path starts from v. The theorem follows from the following claim.

Claim Assume π is an augmenting path w.r.t. M. Then if there is no augmenting path from v w.r.t. M then there is no augmenting path from v w.r.t. $M \oplus \pi$.

We omit an easy proof of the claim.

The thesis now follows from the claim.

□

Similarly as in the *bipartite case* we try to reduce the operation FindPath to finding a path in some auxiliary graph. Let us fix the matching M and vertex v. Denote by A the set of nodes adjacent to any free node different from v.

Construct the auxiliary directed graph $\vec{G}$. Its vertices are the vertices of G.Each edge (p, q) of $\vec{G}$ corresponds to two edges (p, r), (r, q) of G, where $(p, r) \notin M$ and $(r, q) \in M$.

Additionally we include in $\vec{G}$ all edges of G leading from a vertex in A to a free vertex different than v.

Observation There is an augmenting path (w.r.t. M) starting in v iff there is a path π in $\vec{G}$ such that π corresponds to a simple path in the original graph G.

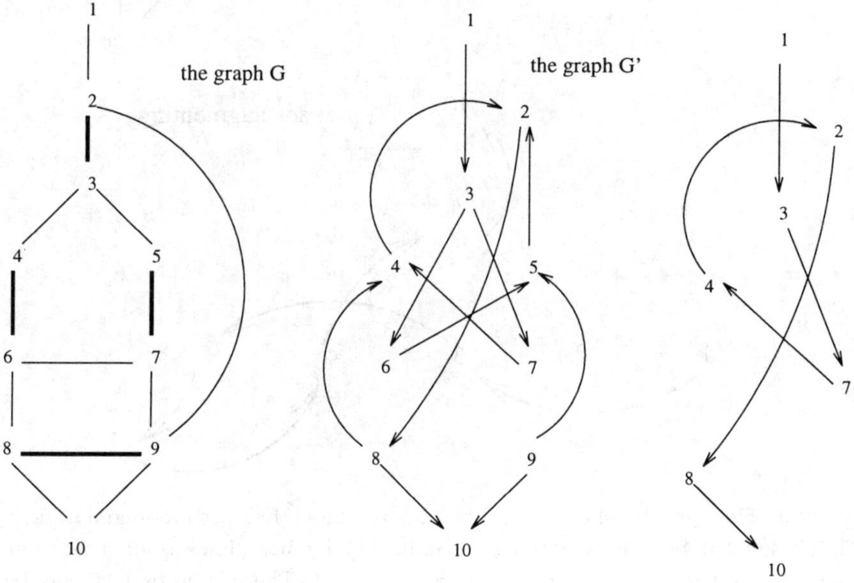

Fig. 3.2 A graph G and its auxiliary graph $G' = \vec{G}$, where $v = 1$. The matching is indicated in bold. We have $A = \{8, 9\}$. The rightmost path is a pseudo-augmenting path but it is not really an augmenting path.

Any path in the graph $\vec{G}$ from v to a free vertex different than v is called a *pseudo-augmenting path*.

Observation It is possible (see Figure 3.2) that G contains a pseudo-augmenting path but has no augmenting path.

If a pseudo-augmenting path does not correspond to a simple path in the original graph then it implies constructively a very special structure called the *blossom*, which is the main concept in sequential algorithms for matchings in general graphs.

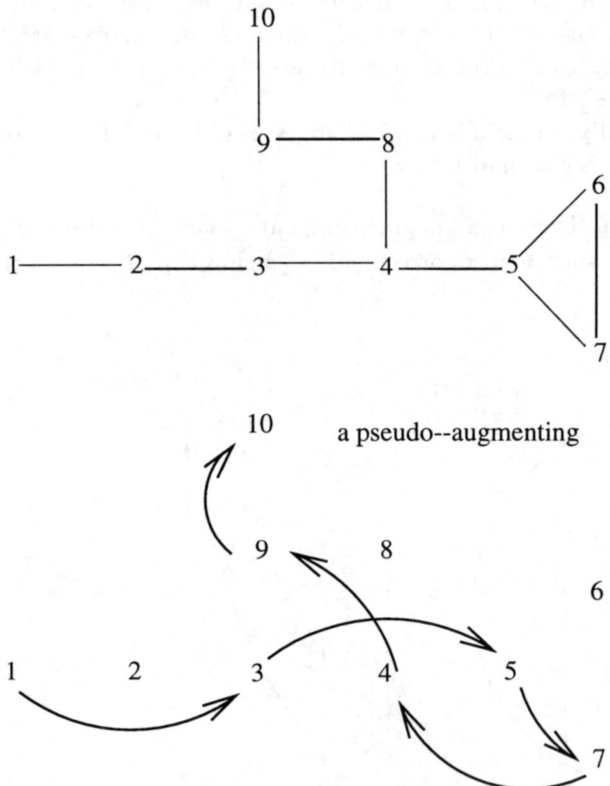

Fig. 3.3 The graph shown in the figure has the pseudo-augmenting path $(1, 3, 5, 7, 4, 9, 10)$ but the matching M indicated by bold lines is of maximum cardinality; hence there is no augmenting path w.r.t. M. There is no perfect matching.

The blossom is an odd cycle B with maximal number of matched edges. If B has $2k+1$ nodes then k edges are matched. The base of the blossom B is the node v which is not matched by the edges of B which are in M. An example of a blossom is illustrated in Figure 3.2 and Figure 3.4 (see the cycle (3,4,6,7,5)).

Assume we have a pseudo-augmenting path γ. This path corresponds to a path π in G. Denote $\pi = \text{full\_path}(\gamma)$.

The path full_path(γ) results from γ by replacing each (*short-cut*) edge (p, q) by edges (p, r), (r, q), where $(r, q) \in M$.

The last edge leading to a free node in A is not replaced, as it does not correspond to a short-cut.

The pseudo-path γ does not correspond to any augmenting path if and only if full_path$(\gamma) = \pi = (v_1, v_2, \ldots, v_k)$ is not a simple path, which means that two vertices repeat on π. In this case let us take the first vertex which repeats on π. This means that we go along the path *on-line* from left to right and catch the shortest prefix (the first situation) $(v_1, \ldots, v_j)$ when the repetition happens (v_j occurs twice in $(v_1, \ldots, v_j)$). Let $(v_i, v_{i+1}, \ldots, v_j)$ be a simple subpath of π which starts and ends with the (first) vertex v_j which repeats. So $v_i = v_j$. Observe that the length of the resulting cycle is odd. Denote

$$\text{blossom}(\gamma) = (v_i, v_{i+1}, \ldots, v_j).$$

For example, if full_path$(\gamma) = \pi = (1, 2, 3, 4, 5, 6, 7, 5, 4, 3, 8)$ then the corresponding blossom (subpath) is blossom$(\gamma) = (5, 6, 7, 5)$, since 5 is the first vertex which repeats (the prefix (1,2,3,4,5,6,7,5) is the shortest prefix with the repetition).

Observe that blossom(γ) is the first blossom in γ, the path γ can contain other blossoms and we can have one blossom inside another blossom. Such nestings of blossoms complicate the matching algorithm.

If there is no blossom corresponding to a pseudo-path then we shall define blossom$(\gamma) = \emptyset$.

Example Consider the graph in Figure 3.2. For a pseudo-augmenting path $\gamma = (1, 3, 7, 4, 2, 8, 10)$ in G' we have

$$\text{full\_path}(\gamma) = (1, 2, 3, 5, 7, 6, 4, 3, 2, 9, 8, 10), \text{and blossom}(\gamma) = (3, 5, 7, 6, 4, 3).$$

Let B be a blossom. Denote by SHRINK(G, B) the graph resulting from G by shrinking all vertices of B in G to a single new vertex $\hat{B}$. This new vertex is adjacent to all vertices in $V - B$ which are adjacent to any node of B. It can happen that B includes the starting vertex; in this case $\hat{B}$ becomes the starting vertex in the new graph.

Assume π is an augmenting path in the graph SHRINK(G, B). Then it is easy to reconstruct from π an augmenting path in G. Denote such an augmenting path by Expand_Blossom(π, B).

If $\pi = (v_1, v_2, \ldots, v_p, B, v_{p+1}, \ldots, v_s)$ then the extension is done by entering the cycle B from v_p and traversing around the cycle to a node in B adjacent to v_{p+1}. The single node $\hat{B}$ is replaced in π by the part of the traversed cycle. We can traverse the cycle in two opposite directions and we choose the correct one (to have an augmenting path).

Example Let us look at Figure 3.2 and Figure 3.4. For a pseudo-augmenting path $\gamma = (1, 3, 7, 4, 2, 8, 10)$ in G' we have: $B = \text{blossom}(\gamma) = (3, 5, 7, 6, 4, 3)$. The graph SHRINK$(G, B)$ contains an augmenting path $\pi = (1, 2, B, 9, 8, 10)$. Then G has an augmenting path

$$\text{Expand\_Blossom}(\pi, B) = (1, 2, 3, 5, 7, 9, 8, 10).$$

The blossom$B$ is replaced by its part $(3, 5, 7)$.

Hence if B is a blossom and SHRINK(G, B) has an augmenting path then G also has an augmenting path. The inverse implication is also true, though harder to prove. We refer to [LP86] and [PS82] for the proof. We summarize the discussion above in the following Lemma.

Lemma 3.2.2
Let B be a blossom in G. Then G has an augmenting path iff $SHRINK(G, B)$ has an augmenting path.

Proof We use our *structural approach* from Chapter 2, in particular the concept of odd-set cover. Recall that an *odd-set cover* of the graph G is any family $\mathcal{F}$ of odd cardinality subsets of V such that each edge of G is incident to a vertex in a one-element set in $\mathcal{F}$ or has both endpoints inside a set belonging to $\mathcal{F}$. The cost of $\mathcal{F}$ is the sum of costs of sets in $\mathcal{F}$, where the cost of a $2k+1$-element set is defined to be $\max\{k, 1\}$. We have shown the following fact (*duality theorem* in Chapter 2): maximum size of a matching in a graph G equals the minimum cost of an odd-set cover.

Assume the blossom B contains $2r + 1$ edges and the actual matching M of the graph G contains k edges. Let $G' =$ SHRINK(G, B) and M' consist of all edges of M which are not contracted to a single node. Hence $|M'| = k - r$. Using the *duality theorem* we show the following claim.

Claim If there is no augmenting path in G' with respect to M' (M' is a maximum cardinality matching in G'), then there is no augmenting path in G with respect to M.

If M' is of maximum cardinality then, according to *duality theorem* there is an odd-set cover $\mathcal{F}'$ whose cost equals $k - r$. Now it is enough to show that there is an odd-set cover $\mathcal{F}$ in G whose cost is k.

Let B' be the node in G' corresponding to the blossom B. The *blossom node* B' is contained in a set $Z \in \mathcal{F}$. We consider two cases:

Case 1: $|Z| > 1$.
Replace Z by $Z - \{B'\} \cup B$. Then the resulting family $\mathcal{F}$ of sets is an odd-set cover of G whose cost is k.

Case 2: $|Z| = 1$.
In this case we add to the family $\mathcal{F}'$ new set B (consisiting of all nodes in the blossom). The resulting family $\mathcal{F}$ is a required odd-set cover of G.

In this way we have shown the harder part (inverse implication): if SHRINK(G, B) has no augmenting path then G also does not have such path.

We have shown before (in a very straightforward way) that if SHRINK(G, B) has an augmenting path then also G has.

This completes the proof of the lemma.

□

We have two basic operations related to blossoms: SHRINK and Expand_Blossom. Assume that Expand_Blossom$(\emptyset, B) = \emptyset$. Correctness of the algorithm follows from the fact that whenever a pseudo-augmenting path γ is *bad* (does not correspond to an augmenting path) then γ contains a blossom. The computation of FindPath(v, M) is the main and most costly part of the algorithm. The main point is that the problem is reduced to $O(n)$ applications of searching a pseudo-augmenting path. This can be done in $O(n^2)$ time by standard methods which look for a path in a *standard* sense (one can forget about the matching for a moment). A DFS or BFS can be used. This implies that the total time of FindPath is $O(n^3)$ and the total time of the whole algorithm is $O(n^4)$.

Theorem 3.2.3
A maximum matching in a general undirected graph can be found in $O(n^4)$ time.

Our version is a simplified version of the Edmonds algorithm. It is less effective than the original algorithm (though simpler to understand).

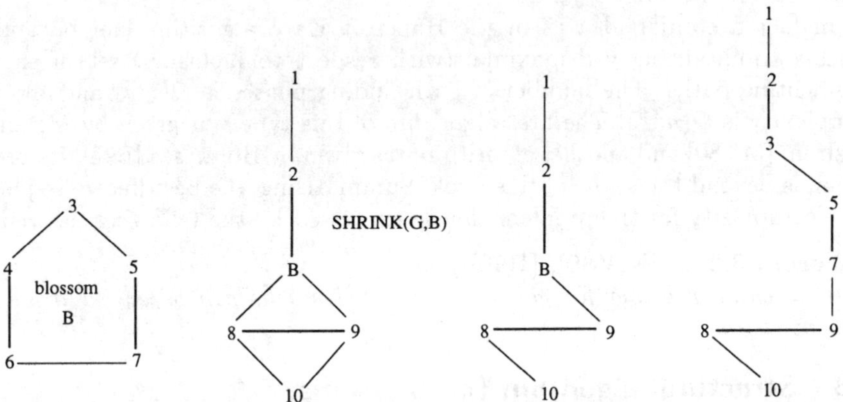

Fig. 3.4 A blossom in the graph from Figure 3.2, the graph SHRINK(G, B), an augmenting path π in SHRINK(G, B) and the path Expand_Blossom(π, B). The last path is the result of FindPath(G).

```
Function FindPath(G);
{returns an augmenting path from the starting node to a node in A,
returns ∅ if there is no such path}
construct the graph G' = G⃗;
find any pseudo-augmenting path γ in G';
if γ = ∅ then return ∅
   else if full_path(γ) is an augmenting path in G
          then return full_path(γ)
   else begin
          B := blossom(γ);
          return Expand_Blossom(FindPath(SHRINK(G, B)), B);
          end
```

In the original Edmonds algorithm, when we search for a pseudo-augmenting path we stop in the first moment when a blossom is discovered. Then the blossom is shrunk and the search continues, the constructed part of a *DFS-tree* (or *BFS-tree*) being preserved. In this way we save some time, since we do not wait to shrink a blossom until a complete pseudo-augmenting path is found. The vertices to be processed can be put on a queue Q similarly to BFS searching. Each time we take a node from Q we process all the edges adjacent to v. A given vertex is processed only once. We omit details and refer to [B91], [PS82] for the description of the Edmonds algorithm.

As was noticed by Blum in [B91] the original Edmonds algorithm works in $O(n^3)$ time. Previously it was believed that it works in (n^4) time, as claimed for example in [LP86], and several authors refined the Edmonds algorithm considerably to get the cubic bound.

In fact a similar idea as in the Hopcroft–Karp algorithm can be applied: process simultaneously a maximal (with respect to inclusion) set of shortest augmenting paths. The number of path finding phases is $O(\sqrt{n})$ and the total complexity is $O(n^{2.5})$. The first algorithm of this type was given by Micali and Vazirani [MV80] and another algorithm was given by Blum, see [B90]. Its presentation is beyond the scope of this book. Summarizing, the best known sequential time complexity for the problem can be expressed by the following theorem.

Theorem 3.2.4 ([MV80], [B90])
The matching problem for general undirected graphs can be solved in $O(n^{2.5})$ sequential time.

3.3 Structural algorithm (no blossoms)

In this section we present yet another sequential polynomial time algorithm which is an algorithmic implementation of the witness theorem from Chapter 2. The algorithm also gives a constructive proof of this theorem. The witness theorem claims that there is always a very special set whose local deficiency is the same as the global deficiency of the whole graph. This special set was denoted in Chapter 2 by TutteSet(G) and it was defined in terms of critical vertices. In this section we refine the notion of critical vertices.

Let $\mathcal{L}$ be a family of (not necessary maximum) matchings of the same cardinality. We say that a vertex $v \in V$ is *$\mathcal{L}$-critical* iff v is *matched* in each matching in $\mathcal{L}$. In other words v is not "missed" by any matching in $\mathcal{L}$. We say that a matching M misses a vertex v, if v is not matched by M.

Observe that if $\mathcal{L}$ is the family of all maximum cardinality matchings then "$\mathcal{L}$-critical" means that after removing v from the graph the matching number decreases.

Define three sets of vertices:

$$\begin{aligned} D(\mathcal{L}) &= \{v \in V\colon\ v \text{ is not } \mathcal{L}\text{-critical}\}. \\ A(\mathcal{L}) &= \{v \in V\colon\ v \text{ is } \mathcal{L}\text{-critical and is adjacent to a vertex in } D(\mathcal{L})\}. \\ C(\mathcal{L}) &= V - (A(\mathcal{L}) \cup D(\mathcal{L})). \end{aligned}$$

The decomposition $V = D(\mathcal{L}) \cup A(\mathcal{L}) \cup D(\mathcal{L})$ is called here the *Gallai–Edmonds decomposition*, see also [LP86].

Fact 3.3.1
If $\mathcal{L}$ is the family of all maximum cardinality matchings of G then $A(\mathcal{L})$ = TutteSet(G) *(as defined in Chapter 2).*

Proof A simple proof is omitted.

□

If $T \subseteq V$ then $G[T]$ denotes the subgraph of G induced by T. Denote by Restricted(M,T) the matching consisting of all edges of a matching M whose endpoints are in T, for a set $T \subseteq V$. In other words Restricted$(M,T) = M \cap E(T)$, where $E(T)$ are the edges of $G[T]$.

We say that that a connected component T of the graph G has a *near-perfect* matching w.r.t. M if all vertices of T except exactly one vertex are matched by the matching Restricted(M,T).

A matching M is said to be *$\mathcal{L}$-good* iff the following conditions hold:

1. There is no edge of M between $A(\mathcal{L})$ and $(A(\mathcal{L}) \cup C(\mathcal{L}))$.
2. Each connected component of $G[D(\mathcal{L})]$ has a near-perfect matching w.r.t. M.

A matching M is said to be *$\mathcal{L}$-bad* iff it is not $\mathcal{L}$-good.

Example Consider the matching M shown in Figure 3.5, where the structure of the graph is shown with respect to some family $\mathcal{L}$. Assume that $M \in \mathcal{L}$ is $\mathcal{L}$-good for the family $\mathcal{L}$, where M is indicated by bold edges. The sets $A(\mathcal{L})$, $D(\mathcal{L})$ and $C(\mathcal{L})$ are indicated as A, D and C. Observe that there is no edge between C and D, there is no matching edge between A and $A \cup C$ and each vertex in A is matched to a vertex in D. The deficiency equals the number of missed vertices in D. We have $|A| = 3$ and $\theta(G - A) = 5$, since there are five odd components in $G[D]$. Hence the local deficiency of A equals $5 - 3 = 2$. Each component of $G[D]$ has a near-perfect matching w.r.t. M. The set A is a witness set.

Recall that the set

$$X \text{ is a witness set of } G \text{ iff } \text{DEFECT}(G) = \text{LOCAL\_DEFECT}_G(X),$$

where $\text{LOCAL\_DEFECT}_G(X) = \theta(G - X) - |X|$ and θ means "the number of odd connected components".

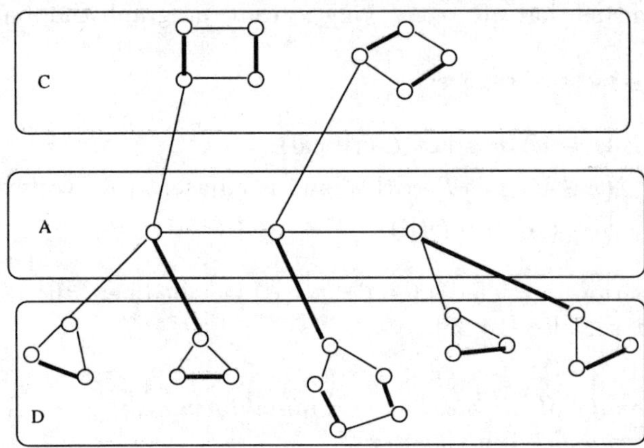

Fig. 3.5 The structure of an example graph; the edges of a matching M from the family $\mathcal{L}$ are indicated in bold.

Lemma 3.3.2
Let $\mathcal{L}$ be a family of matchings of the same cardinality k of the graph G and $M \in \mathcal{L}$. Then if M is $\mathcal{L}$-good then $A(\mathcal{L})$ is a witness set of G and M is a maximum cardinality matching.

Proof All components of $D(\mathcal{L})$ are odd, since they have near-perfect matchings. On the other hand all components of $C(\mathcal{L})$ are even, because they have perfect matchings. Hence $\theta(G - A(\mathcal{L}))$ equals the number p of connected components of $D(\mathcal{L})$.

For each vertex v in $A(\mathcal{L})$ there is an edge in M connecting v to a vertex in a component in $D(\mathcal{L})$. The number of missed vertices equals $(p - |A(\mathcal{L})|)$, which is the local deficiency of the set $A(\mathcal{L})$. Hence $A(\mathcal{L})$ is a witness set and M is a maximum cardinality matching.

□

Theorem 3.3.3
Let $\mathcal{L}$ be a family of matchings of same cardinality k of the graph G. Assume there is a matching $M \in \mathcal{L}$ which is $\mathcal{L}$-bad. Then there is a matching $M' =$ NextMatching($\mathcal{L}$) satisfying one of the following conditions:

1. $|M'| = k + 1$.

2. $|M'| = k$, *and* $|D(\mathcal{L} \cup \{M'\})| = |D(\mathcal{L})| + 1$.
   (In other words M' *is missing a vertex in* $A(\mathcal{L}) \cup C(\mathcal{L})$*.)*

We postpone the algorithmic proof but describe how NextMatching is computed. First we consider the following important theoretical consequence of Theorem 3.3.3: the validity of Theorem 2.4.5. Assume $\mathcal{L}$ is the set of all maximum cardinality matchings; then obviously there is no matching NextMatching($\mathcal{L}$). Hence all matchings in $\mathcal{L}$ are $\mathcal{L}$-good. According to Fact 3.3.1 $\mathcal{L}$ consists of all maximum cardinality matchings and $A(\mathcal{L})$ = TutteSet(G). This implies the following.

Corollary 3.3.4
Theorem 2.4.5 is valid: TutteSet(G) *is a witness set of the graph* G.

Another important consequence is the construction of a new algorithm. The general structure of an algorithm computing a maximum cardinality matching by using the operation NextMatching (which is described later) is as follows.

```
Structural algorithm;
{the algorithm returns a maximum cardinality matching
together with a witness set};
M := any initial nonempty matching;
k := |M|; L := {M};
while M is L-bad do
    begin
        M := NextMatching(L);
        if |M| > k then L := {M}
           else L := L ∪ {M};
        k := |M|;
    end
return maximum cardinality matching M
       and the witness set A(L).
```

In the description of the operation NextMatching the crucial role is the operation of *releasing* a vertex v. If v was matched w.r.t. a given matching then after the operation it is missed by a newly created matching.

Descriptions of operations releasing a given vertex Assume π is an M-alternating path w.r.t. a given matching M, $v \in \pi$, π ends with a vertex w which is missed by M and the distance on π from v to w is even. Then we can *switch* the matching on the path π (the size of the matching does not change), but after the operation v becomes a missed vertex. The operation is denoted by ReleaseOnPath(v, π, M) and is illustrated schematically in Figure 3.6.The value of the operation is the newly constructed matching.

We define a similar operation for cycles. Let C be an odd alternating cycle w.r.t. M with one missed vertex w. Then the operation ReleaseOnCycle(v, C, M)

changes the matching without changing its size, so that in the new matching v is missed. The value of the operation is the newly constructed matching.

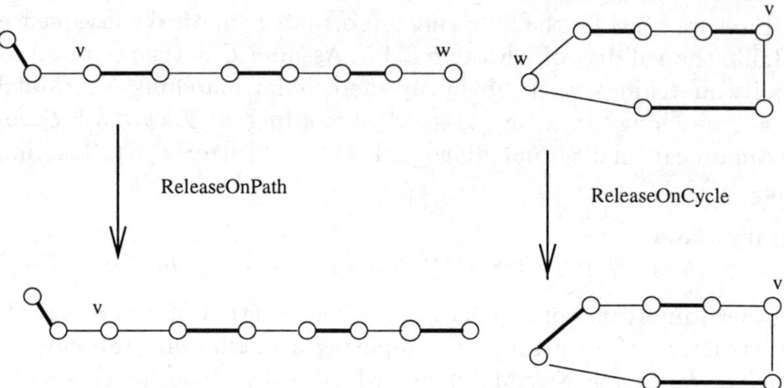

Fig. 3.6 The operations ReleaseOnPath(v, π, M) and ReleaseOnCycle(v, C, M). The matched edges are in bold. The distance from v to w on the path π is even. Initially v is matched and w is missed. Then the situation is reversed.

Introduce also a useful operation AltPath(v, M_1, M_2), where $v \in V$, M_1, M_2 are matchings and v is missed by M_2 and not by M_1. The value of the operation is the inclusion maximal (unique) path starting at v and consisting of edges in $M_1 \oplus M_2$. We shall frequently use the following fact.

Fact 3.3.5
If v is missed by M_2 and $\pi =$ AltPath(v, M_1, M_2) ends with an edge in M_1 then π is an augmenting path w.r.t. M_2.

Let us fix a family $\mathcal{L}$ of k-element (partial) matchings of G, for some fixed k. For simplicity of presentation we write D, A, C for (respectively) $D(\mathcal{L})$, $A(\mathcal{L})$ and $C(\mathcal{L})$.

Assume M' is a k-element matching (M' is *not necessarily* a member of $\mathcal{L}$).

We introduce an auxiliary operation AUX(x, y, M'). The operation works under the following assumption

$$\text{COND}(x, y, M'):$$

(1) there is a connected component T of D,
(2) there are two vertices $x, y \in T$ which are missed by Restricted(M', T),
(3) x is also missed by M'.

The value of AUX(x, y, M') is a matching M which misses a vertex in A or whose cardinality is $k+1$. This matching will be also used as the value of NextMatching.

DESCRIPTION OF THE OPERATION AUX(x, y, M'):

Case 1: x, y are adjacent.
If y is missed by M' then **return** $M' \cup \{(x, y)\}$
else $(y, v) \in M'$ for some $v \in A$, **return** $(M' - \{(y, v)\}) \cup \{(x, y)\}$.

Case 2: x, y are not adjacent.
Let $z \in T$ be a vertex which follows x on the shortest path from x to y in the component T. Let M'' be a matching in $\mathcal{L}$ missing vertex z.

Subcase 2.1: z is missed by M' (such a matching exists by the definition of D): **return** $M' \cup \{(x, z)\}$.

Subcase 2.2: z is not missed by M' and the path $\pi = \text{AltPath}(z, M', M'')$ does not end in x, where M'' is a matching in $\mathcal{L}$ *missing* z.

In this case π is an augmenting path w.r.t. M''. Using π as an augmenting path we construct a matching of size $k + 1$ and return it as the value of AUX(x, y, M'). The operation is completed.

Subcase 2.3: z is not missed by M' and the path $\pi = \text{AltPath}(z, M', M'')$ ends in x and contains y.

The path π together with the edge (x, z) becomes an alternating cycle C w.r.t. M'. The vertex x is missed by M' and there is some vertex $v \in A$ which is joined with y by an edge of M'. Notice that $v \in C$. Then we return the matching ReleaseOnCycle(v, C, M'). The newly constructed matching is missing a vertex $v \in A$. The size of D increases and the operation AUX is completed.

Subcase 2.4: z is not missed by M' and the path $\pi = \text{AltPath}(z, M', M'')$ ends in x and does not contain y, see Figure 3.7.

Let $M' = \text{ReleaseOnPath}(z, \pi, M')$. Then **return** $AUX(z, y, M')$.

The distance from z to y is smaller than that of x from y. After several iterations we arrive at Case 1 or at one of the previously described cases. This completes the description of the operation AUX.

DESCRIPTION OF THE OPERATION NextMatching$(\mathcal{L})$:

Let $M \in \mathcal{L}$ be a matching which is $\mathcal{L}$-bad.

Case I: there is a connected component T of D, with two nodes $x, y \in T$ such that x is missed by M and y is missed by Restricted(M, T). In this case **return** $AUX(x, y, M)$.

Case II: there is a connected component T of D with two vertices $x, y \in T$ which are missed by Restricted(M, T) and none of them is missed by M.

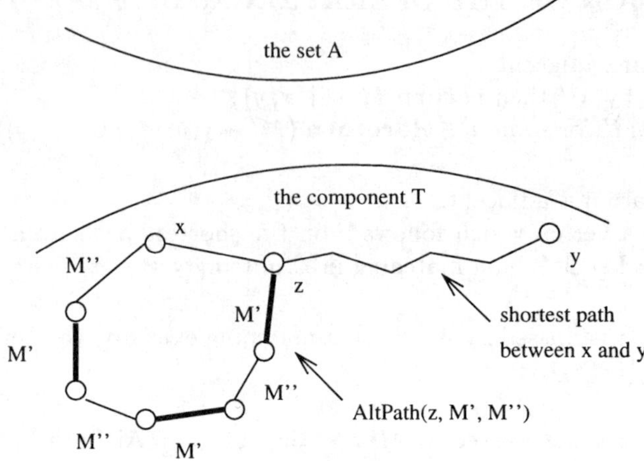

Fig. 3.7 Subcase 2.4 in the description of AUX(x, y, M'). The vertex x is missed by M' and y is missed by Restricted(M', T). z is the vertex following x on the shortest path to y. There is a matching $M'' \in \mathcal{L}$ missing z. The matching M' is "switched" on the path AltPath(z, M', M'') and the role of x is played now by z, which is closer to y. Now COND(z, y, M') holds, where M' is a modified matching (of the same size).

Hence x, y are matched, respectively, with some nodes $u, v \in A$. Let M_1 be a matching in $\mathcal{L}$ missing the vertex x. Take $\pi =$ AltPath(x, M, M_1). Observe that π is not a cycle.

Subcase II.1: if the path π ends with an edge in M, then it is an augmenting path w.r.t. M_1, according to Fact 3.3.5. Hence we can produce a matching of cardinality $k + 1$. This matching is returned as the value of the whole operation.

Subcase II.2: the path π ends with an edge in M_1 and does not contain y; then **return** AUX$(x, y,$ ReleaseOnPath$(x, \pi, M))$.

Subcase II.3: the path π ends with an edge in M_1, contains y and y is at an even distance from the end of the path (which is *missed* by M). Then **return** AUX$(y, x,$ ReleaseOnPath$(y, \pi, M))$.

Subcase II.4: the path π ends with an edge in M_1, contains y and y is at an odd distance from the unmatched (by M) end of the path. In this case a neighbor $v \in A$ of y is joined with y by an edge of the path π. This neighbor is at an even distance from the end of the path. Take $M_2 =$ ReleaseOnPath(v, π, M). M_2 is missing a vertex in A and has cardinality k. Then **return** M_2.

Case III: there is a component T of D possessing a perfect matching w.r.t. M.

Then $|T|$ is even. There is some other matching M_1 such that Restricted(M, T) is missing a vertex in T; then (due to an even number of nodes of T) it misses at least two vertices. Then we are at one of the previous cases.

Case IV: M contains an edge (x, y) between A and $A \cup C$.

By the definition of the set A there is some edge (x, z) for a vertex $z \in D$. There is also a matching $M' \in \mathcal{L}$ missing the vertex z.

Subcase IV.1: M is missing z. Then **return** $M - \{(x, y)\} \cup \{(x, z)\}$.

Subcase IV.2: let $\pi =$ AltPath(z, M, M'). If π ends with an edge of M then π is an augmenting path w.r.t. M'. We can produce a matching of cardinality $k + 1$.

Subcase IV.3: π ends with an edge of M' and does not contain (x, y), see Figure 3.8.

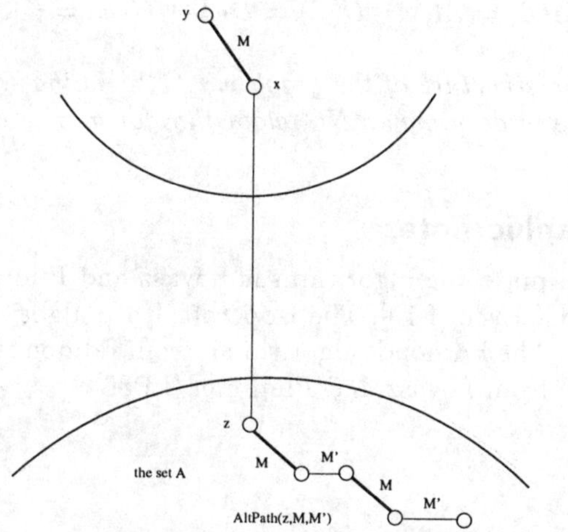

Fig. 3.8 A schematic illustration of Subcase IV.3 in the description of NextMatching$(\mathcal{L})$. There is an edge of M between a vertex $x \in A$ and a vertex $y \in A \cup C$. We can switch on the path shown in the figure and release a vertex in $A \cup C$.

We can use the path $\pi' = (y, x, z, \pi)$. Then **return** ReleaseOnPath(y, π', M).

Subcase IV.4: π ends with an edge of M' and contains (x, y). Then one of the vertices x or y is at an even distance from the end of π. We use the operation ReleaseOnPath to produce a matching M_2 which misses x or y. Then **return** M_2.

Theorem 3.3.6

The structural algorithm constructs a maximum cardinality matching and a witness set in $O(n^4)$ time. It can also report (within the same complexity) the

Gallai–Edmonds decomposition $V = D(\mathcal{M}) \cup A(\mathcal{M}) \cup D(\mathcal{M})$, *where* $\mathcal{M}$ *is the family of all maximum cardinality matchings.*

Proof There are at most n^2 iterations (and also applications of the operation NextMatching) in the algorithm, since, in each iteration we increase the size of the matching or the size of the set $D(\mathcal{L})$. The operation NextMatching works in $O(n^2)$ time, since it involves at most $O(n)$ AltPath operations. These operations are the most costly. Each operation AltPath can be done in linear time. Altogether we have $O(n^4)$ time complexity.

□

Remark 3.3.7
Another interesting consequence of the structural algorithm is that there is a family $\mathcal{L}$ *of linearly many maximum cardinality matchings which in the following sense represents the whole (usually of exponential size) family* $\mathcal{M}$ *of all maximum cardinality matchings:*

$$A(\mathcal{L}) = A(\mathcal{M}), D(\mathcal{L}) = D(\mathcal{M}), C(\mathcal{L}) = C(\mathcal{M}).$$

In other words the structure of the graph w.r.t $\mathcal{M}$ *is the same as w.r.t.* $\mathcal{L}$.

Is it of any use in deriving an NC-algorithm for maximum cardinality matchings?

3.4 Bibliographic notes

The basic text on matching algorithms is Lovász and Plummer [LP86], see also Gibbons [Gi85] and Even [E79]. The Hopcroft–Karp algorithm is from Hopcroft and Karp [HK73]. The Edmonds algorithm is from Edmonds [E65] and the structural algorithm is from Lovász and Plummer [LP86].

4

Probabilistic tools

4.1 Probabilistic tools

In this book we use several (rather simple) facts from probability theory. We state and prove most of them in this chapter and they will be used later to justify the effectiveness of our randomized algorithms.

Now we introduce some very basic terminology and formalism. For more details refer to standard textbooks on probability theory, for example [Ch74].

A *probability space* is a triple $(\Omega, \mathcal{F}, Pr)$, where

Ω is a set (sample space);

$\mathcal{F}$ is a collection of subsets of Ω (with $\Omega \in \mathcal{F}$ and closed under the complement and union). The elements of $\mathcal{F}$ are called *probabilistic events*;

Pr is the probability measure on $\mathcal{F}$, with

$$Pr(\Omega) = 1 \quad \text{and} \quad Pr(A \cup B) = Pr(A) + Pr(B) \quad \text{if} \quad A \cap B = \emptyset.$$

Several assumptions should be made about the structure of $\mathcal{F}$, but in this book we use rather "simple" probabilistic spaces.

We assume (in this book) that Ω is a *discrete* finite or infinite enumerable set

$$\Omega = \{\omega_1, \omega_2, \ldots, \omega_r, \ldots\},$$

and probabilistic events are all subsets of Ω. The probability measure is defined for $Z \subseteq \Omega$ as

$$Pr(Z) = \sum_{\omega_i \in Z} Pr(\omega_i).$$

In many cases the probability of an event is described concisely in the form

$$Pr\{\textit{description of an event}\}.$$

For example, if the probability space consists of all binary sequences of an even length n, each one with the same probability, then we can describe

$$Pr\{\textit{the number of ones in the sequence is odd}\} = \frac{1}{2}.$$

Independent repetition is a powerful trick. Assume we try randomly to find some object X and the probability that we find it is at least $1/2$. Assume that we repeat this process, and each stage is probabilistically independent of each other (*failure* consists of not finding X).

The probability that all 25 trials are failures does not exceed 0.00000003. This probability is so small that we could optimistically treat the whole computation as a deterministic one (always terminating after at most 25 trials with *success*). Failure is unlikely; in real life we could expect many other failures with such small probability, but we do not take them seriously.

A *random variable* is a function $X\colon \Omega \rightarrow \mathcal{R}$ (from the elements of the probabilistic space to real numbers). The expected value of the random variable X is the value

$$E[X] = \sum_{\omega \in \Omega} Pr(\omega) \cdot X(\omega).$$

For example, let X be the number of ones in a random binary sequence of length n. Then

$$E[x] = \sum_{k=1}^{n} k \cdot \binom{n}{k} \cdot 1/2^n.$$

An important expected value considered in this book is the expected time of a randomized algorithm and the most important property of the expected value that we shall use is that it is additive:

$$E[X+Y] = E[X] + E[Y] \quad \text{and} \quad E[c \cdot X] = c \cdot E[X] \quad \text{for a constant } c.$$

4.2 The isolating theorem

In this and the next section we present two results related to families $\mathcal{M}$ of sets which are rather surprising in view of their generality. These results are general, but we can keep in mind in the next chapters that our basic families of subsets $\mathcal{M}$ will be families of all maximum matchings for given graphs.

We shall assign integer weights to elements of a set $A = \{a_1, a_2, \ldots, a_m\}$. The set of weights is denoted by $\mathcal{W}$, and we assume it consists of a finite number of consecutive integers.

Define the weight of a subset as the sum of weights of its elements. Let $\mathcal{M}$ be a family of subsets of a set A and f a function assigning integer weights to elements of A. We say that f has the *uniqueness property* iff there is only one set in $\mathcal{M}$ of minimal weight with respect to f.

Our aim is to show that if $\mathcal{W}$ is sufficiently large and we assign weights from $\mathcal{W}$ in random then the assignment f has uniqueness property with high probability.

Example Consider the following family of subsets of $A = \{1, 2, 3, 4\}$:

$$\mathcal{M} = \{\{1,2\}, \{1,3\}, \{2,4\}, \{3,4\}, \{2,3\}, \{1,4\}\}$$

$\mathcal{M}$ consists of all 2-element subsets of A. Assume that the set of weights is

$$\mathcal{W} = \{1,2,3,4\}.$$

In this case the assignment $f = (1,2,1,1)$ has no uniqueness property, since we have three minimal sets in $\mathcal{M}$: {1,3}, {3,4} and {1,4}. On the other hand the assignment $f = (1,2,1,2)$ has the uniqueness property.

In this particular case we shall compute exactly the probability that randomly chosen assignment f has the uniqueness property.

We have the following property: an assignment f has the uniqueness property iff there are two elements $a_i, a_j \in A$ such that

$$f(a_k) > f(a_i) \text{ and } f(a_k) > f(a_j) \text{ if } k \neq i \text{ and } k \neq j.$$

These elements a_i, a_j have the first and the second minimal weights.

We have the following six possibilities for the first and second minimal weight:

- (1,1): in this case the other two weights are from the set {2,3,4}, and hence we have 3^2 possibilities for other weights if elements having weights 1, 1 are fixed. All together there are $\binom{4}{2}3^2 = 6 \cdot 9 = 54$ possibilities, since any 2 elements of A can have weights 1 and 1.
- (1,2): two other weights are in {3,4}. We have altogether 48 possibilities, since we can choose elements having weights 1 and 2 in 12 ways.
- (1,3): we have 12 possibilities.
- (2,2): we have 24 possibilities.
- (2,3): we have 12 possibilities.
- (3,3): we have 6 possibilities.

Altogether there are 54+48+12+24+12+6 = 156 assignments f having the uniqueness property, out of $4^4 = 256$ all possibilities. Hence in our case:

> probability that a random assignment has the uniqueness property equals $\frac{156}{256} > \frac{1}{2}$.

Assume that a function

$$\phi\colon (A - \{a_i\}) \to \mathcal{W}$$

is an assignment of values from a set $\mathcal{W}$ of integers to all elements of A except a_i. We also denote (informally) the function ϕ as the assignment vector

$$\phi = (w_1, w_2, \ldots, w_{i-1}, *, w_{i+1}, \ldots, w_m).$$

Such a vector means that $\phi(a_j) = w_j$, for $j \neq i$, and the weight $\phi(a_i)$ is not defined.

We say that $\theta = \text{threshold}(\phi) = \text{threshold}(w_1, w_2, \ldots, w_{i-1}, *, w_{i+1}, \ldots, w_m)$ is a *threshold value* for ϕ iff one of the following conditions holds after assigning to a_i any new weight $w_i \neq \theta$:

(1) a_i is in *each set* from $\mathcal{M}$ of minimal weight, or

(2) a_i is in *no set* from $\mathcal{M}$ of minimal weight.

If there are many threshold values possible then we choose the smallest one.

The i-th element a_i is said to be *ambiguous* with respect to an assignment $f = (w_1, w_2, \ldots, w_m)$ iff $w_i = \text{threshold}(w_1, w_2, \ldots, w_{i-1}, *, w_{i+1}, \ldots, w_m)$.

Example Consider again the family

$$\mathcal{M} = \{\{1,2\},\ \{1,3\},\ \{2,4\},\ \{3,4\},\ \{2,3\},\ \{1,4\}\}$$

Let the set of weights be $\mathcal{W} = \{1,2,3,4,5,6\}$. Then $\text{threshold}(3,*,5,3) = 3$, since for the assignment (3,3,5,3) the second element 2 is in the minimal subset {1,2} and is not in the minimal subset {1,4}. In other words the element 2 is ambiguous with respect to the assignment (3,3,5,3).

If the weight of 2 is larger than 3 then 2 is in no minimal set of $\mathcal{M}$; if it is smaller then it is in all minimal sets of $\mathcal{M}$.

Lemma 4.2.1
(A) *If the assignment f does not have the uniqueness property then there is an ambiguous element with respect to f.*
(B) *For each partial assignment $(w_1, w_2, \ldots, w_{i-1}, *, w_{i+1}, \ldots, w_m)$ there exists a threshold value.*

Proof The point (A) is rather obvious, assuming the existence of threshold values. Let us go to the proof of point (B) and show the existence of threshold values.

Take a value θ satisfying the condition:

(*) there exists a set containing a_i and there exists a set not containing a_i, both having minimal weight with respect to the assignment

$$f = (w_1, w_2, \ldots, w_{i-1}, \theta, w_{i+1}, \ldots, w_m).$$

It is easy to see that such value of θ satisfies properties of $\text{threshold}(\phi)$. This proves the existence of a threshold, in cases when such θ exists.

If there is no θ satisfying the condition (*) then any value of θ is good as a threshold value.

□

The following theorem is quite surprising in view of its generality; it was shown in [MVV87].

Theorem 4.2.2 (isolating theorem)
Assume $\mathcal{M}$ is a family of subsets of a certain set $\mathcal{A}$ of cardinality m. Assume we assign to each element $a_i \in \mathcal{A}$ a random integer $f(a_i) \in [1 \ldots 2m]$. Then

$$Pr\{f \text{ has the uniqueness property}\} \geq \frac{1}{2}.$$

Proof Let f be a random assignment of weights (elements of $[1 \ldots 2m]$) to elements of $\mathcal{A} = \{a_1, \ldots, a_m\}$. Consider a given element $a_i \in \mathcal{A}$. We claim the following:

$$Pr\{a_i \text{ is ambiguous}\} \leq 1/2m.$$

The claim above follows from the fact that after specifying a partial assignment

$$\phi = (w_1, w_2, \ldots, w_{i-1}, *, w_{i+1}, \ldots, w_m)$$

the element a_i can be ambiguous only if $w_i = \text{threshold}(\phi)$.

Hence only one value of the weight for a_i causes the ambiguity of a_i. However, $2m$ values are possible for the weight of a_i. So the probability that a_i is ambiguous is at most $1/2m$.

There are m elements a_i. So the total probability that one of them is ambiguous is at most

$$m \cdot \frac{1}{2m} = \frac{1}{2}.$$

Hence the probability that f has the uniqueness property is at least $\frac{1}{2}$.

□

4.3 The redundancy theorem

Let $\mathcal{M}$ be a family of subsets of a given set E of m elements. In the next chapters E will be the set of edges of a certain graph G, and $\mathcal{M}$ the family of maximum matchings.

For $S \subseteq E$ define

$$\begin{aligned} \text{rank}(S) &= \max\{|S \cap M| : M \in \mathcal{M}\} \\ \text{redundant}(S) &= \{e \in E - S : \text{rank}(S) = \text{rank}(S \cup \{e\})\}. \end{aligned}$$

The elements of redundant(S) are called redundant with respect to S. The concept of redundancy is used later in the book to identify one of the subsets of $\mathcal{M}$; its usefulness comes from the following obvious observation.

Observation (main property of redundant elements) Assume the family $\mathcal{M}$ of subsets of E is nonempty, and $S \subseteq E$. Then there is a subset $M \in \mathcal{M}$ such that $M \subseteq E - \text{redundant}(S)$.

Example Figure 4.1 presents the family

$$\mathcal{M} = \{\{1,2,3\}, \{4,5,6\}, \{4,8,10\}, \{7,8,9\}\}$$

of subsets of $E = [1 \ldots 12]$. In this case we have

$$\text{redundant}(\{3,4,7,8\} = \{1,2,5,6,11,12\}.$$

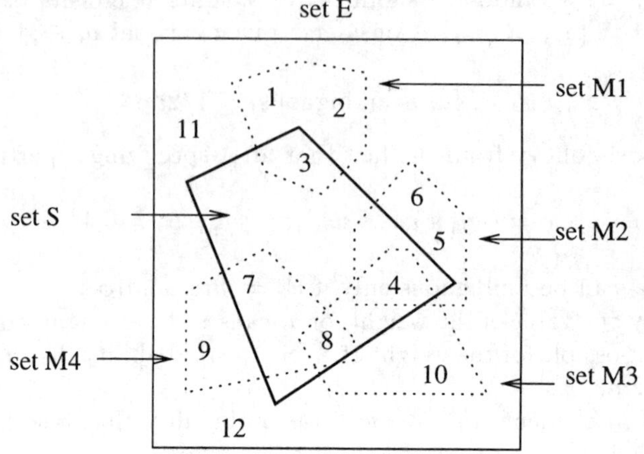

Fig. 4.1 The family $\mathcal{M}$ consists of sets in the dotted lines. Let $S = \{3, 4, 7, 8\}$. Then $\text{rank}(S) = |S \cap M3| = 2$ and $\text{redundant}(S) = \{1, 2, 5, 6, 11, 12\} = |S \cap M4|$.

Observe also that

$$\text{redundant}(\{3, 4, 7\}) = \{11, 12\}, \text{redundant}(\{11, 12\}) = \{1, 2, \ldots, 10\}.$$

Let us look at one more example.

Example Take $\mathcal{M} = \{\{1, 2\}, \{3, 4\}, \{5, 6\}\}$ and consider subsets $S \subseteq [1 \ldots 6]$ of size 3. Then for 12 possible values of S (those containing one of the subsets in $\mathcal{M}$) the number of redundant elements is three. For the other eight possibilities for S we have zero redundant elements. Hence in this case

$$Pr\{|\text{redundant}(S)| \geq 3\} = \frac{12}{20}.$$

Lemma 4.3.1
Let Z be a random variable, $m = \max(Z)$, and $0 \leq \alpha < 1$. Then

$$(*) \quad E(Z) \geq \alpha \cdot m \quad \textit{implies} \quad Pr\left\{Z > \frac{\alpha}{2} m\right\} > \frac{\alpha}{2}.$$

Proof Observe that

$$\alpha \cdot m \leq E(Z) \leq \frac{\alpha}{2} m \cdot Pr\left\{Z \leq \frac{\alpha}{2} m\right\} + m \cdot Pr\left\{Z > \frac{\alpha}{2} m\right\}.$$

Hence

$$\alpha < \frac{\alpha}{2} + Pr\left\{Z > \frac{\alpha}{2} m\right\}.$$

This implies that

$$Pr\left\{Z > \frac{\alpha}{2}m\right\} > \frac{\alpha}{2}.$$

□

We use the lemma to show that if we choose a subset S randomly then we get many redundant elements with high probability. This result was shown in [KUW86].

Theorem 4.3.2 (redundancy theorem)
Let $\mathcal{M}$ be a family of subsets of a set E of size m and k be the maximal cardinality of a set in $\mathcal{M}$. Assume $k \leq \frac{2}{3}m$ and X is a random subset of E. Then

$$Pr\left\{|redundant(X)| \geq \frac{m}{72}\right\} \geq \frac{1}{72}.$$

Proof Take a random subset $X = \{a_1, a_2, \ldots, a_r\}$ of size r, where r is also taken randomly from $[1 \ldots m]$. We can think of X as being constructed by choosing elements $a_1, a_2, \ldots, a_r$ at random.

Let $X_i = \{a_1, a_2, \ldots, a_i\}$, $X_0 = \emptyset$, and let p_i be the probability that randomly taken element $e \in E - X_i$ is *not* in redundant(X_i).

Denote by R_i the expected number of redundant elements with respect to a random set of cardinality i and by R the expected number of redundant elements with respect to any random subset of E. Then

1. $R_i = (1 - p_i)(m - i)$;
2. $\sum_{i=0}^{m-1} p_i = E(\text{rank}(E)) = \text{rank}(E) = k$;
3. $R = \frac{1}{m}\sum_i R_i$.

Claim $R \geq m/36$.

Proof **(of the claim)**

$$R \geq \frac{1}{m}\sum_{i=1}^{\frac{5}{6}m}(1 - p_i)(m - i) \geq \frac{1}{m}\sum_{i=1}^{\frac{5}{6}m}(1 - p_i)\frac{m}{6}$$

$$\geq \frac{1}{6}\left(\sum_{i=1}^{5/6m} 1 - \sum_{i=1}^{m-1} p_i\right) = \frac{1}{6}\left(\frac{5m}{6} - k\right) \geq \frac{m}{6}\left(\frac{5}{6} - \frac{2}{3}\right) = \frac{m}{36}.$$

□

Now we can apply Lemma 4.3.1 with the random variable $Z = |\text{redundant}(X)|$ and $\alpha = \frac{1}{36}$. The coefficient is $\frac{1}{2} \cdot \frac{1}{36} = \frac{1}{72}$. This completes the proof.

□

4.4 Randomized handshaking

Assume we are to choose a large set of disjoint edges of the graph. Imagine the following scenario. There are n persons, with a symmetric relation "A knows B" (such a relation corresponds directly to the edges of the graph). A *mutual handshake* happens if two people A and B who know each other offer their hands. There are two possibilities to create randomly a large set of *mutual handshakes*: first we consider the strategy called *one-phase handshaking.*

One-phase handshaking
Each person gives a hand randomly to a person he (or she) knows.

The one-phase strategy works well for graphs of degree at most two. There are at least n edges in a graph corresponding to the relation "to know each other". Each edge is chosen as a mutual handshake with a probability of at least $1/4$.

This proves the following lemma.

Lemma 4.4.1 (one-phase handshaking)
If each person knows at most two and at least one person then the expected number of mutual handshakes is at least $n/4$.

At first glance it is strange that one-phase handshaking works poorly if there are too many edges. Assume for example that G is the complete graph K_{n+1}. Then the probability that a given edge is selected as a mutual handshake is $1/n^2$, since both endpoints of this edge choose it with the same (independent) probability $1/n$. We have $\binom{n+1}{2} < n^2$ edges, which implies the following.

Observation If we use one-stage handshaking for a complete graph K_n of connections, then the average number of mutual handshakes is less than one mutual handshake.

The second strategy is a refinement of one-phase handshaking. It is more clever, in that two phases are used.

Two-phase handshaking
First phase Each person gives a hand randomly to a person he or she knows.
Second phase Then each person gives a second hand to a randomly chosen person among those who gave a hand to him or her.

Example Assume that we replace *randomness* by *choosing the smaller neighbor.* Then Figure 4.2 illustrates an example of two-phase handshaking. The final set of edges is not necessarily a matching.

Observation Let G' be the graph formed by all mutual handshakes in two-phase handshaking. Then $\deg(G) \leq 2$.

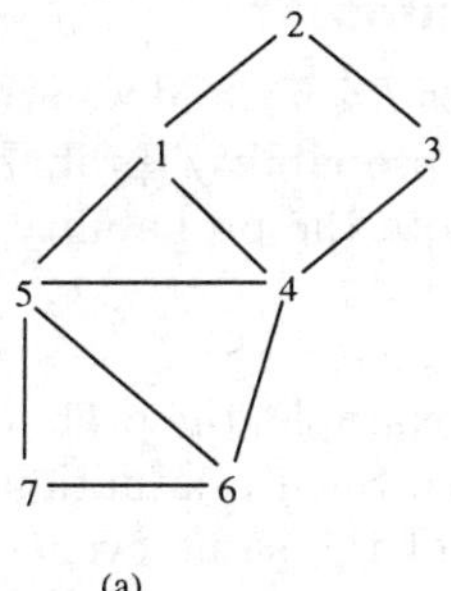

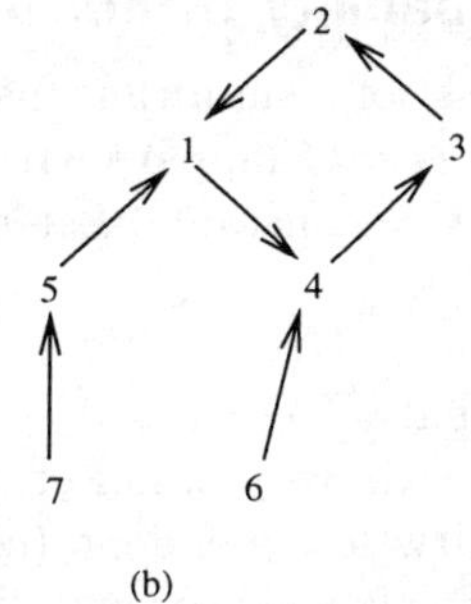

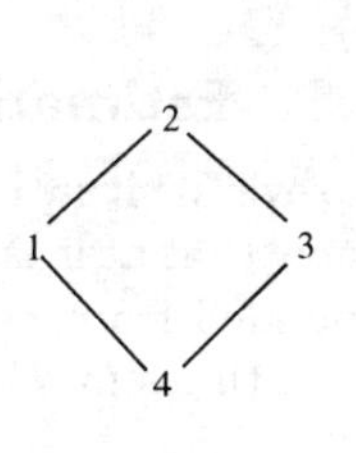

Fig. 4.2 Assume that smaller neighbors are of higher priority, **(a)** a graph G, **(b)** the directed edges selected in the first phase of two-phase handshaking, **(c)** second phase: each node gives a hand to the smallest neighbor who gave a hand in the first phase. The finally selected edges corresponding to mutual handshakes are illustrated.

Proof One person has only two hands.

□

Two-phase handshaking is more clever. We need the following simple fact to see why this strategy is better than one-phase handshaking.

Lemma 4.4.2
Assume $k \geq 1$ is an integer. Then $(1 - 1/k)^k \leq 1/e$.

Proof We omit an easy proof.

□

Observation If we apply two-phase handshaking to the complete graph K_{n+1} then the probability that a fixed node v receives a hand from somebody is

$$1 - (1 - 1/n)^n > 1 - e^{-1} > 1/2.$$

Hence on average at least $n/2$ mutual handshakes are done in K_{n+1}.

The graph corresponding to mutual handshakes has maximum degree 1 or 2. If we now apply one-phase handshaking to such a graph then the resulting graph is of degree at most one. The set of handshakes in the last phase forms a matching of the graph.

If a node had a handshake in the first stage, then this node will have a mutual handshake in the second stage with probability at least $1/2$. Hence the expected number of edges in the resulting matching is at least $n/4$.

In Chapter 5 we apply the randomized handshaking technique to general graphs.

4.5 Estimating the probability of nonzero sums

In several situations we have a set of n simultaneous events $Event_i$ and we need to know that at least one event of $Event_i$ happens with high probability (probability bounded from below by a positive constant). Let p_i denote the probability that the i-th event $Event_i$ happens.

Observation It can happen that $\sum p_i > c > 0$ but, for example, the probability that at least one event happens is very small (e.g. $1/n$). Such a situation can take place if all events are pairwise dependent (copies of the same event) and have the same probability $1/n$. Then the probability that any event of $Event_i$'s happens is $1/n$. If the events are pairwise independent then the above situation is not possible.

Theorem 4.5.1
Assume that the events $Event_i$ are pairwise independent and $\sum p_i > c > 0$. Then

$$Pr\{\text{at least one event of } Event_i \text{ happens}\} \geq c' > 0,$$

where $c' = \frac{2}{3} \cdot \min\{c, 1/6\}$.

Proof If one of p_i is at least $1/6$ then the theorem obviously holds with $c' = 1/6$. So assume that $p_i < 1/6$ for each i. Let us take a subfamily $\mathcal{F}$ of events $Event_i$ such that

$$\min\{c, 1/6\} \leq \sum_{i \in \mathcal{F}} p_i \leq 1/3.$$

Such a family exists since $p_i < 1/6$ for each i.

It is enough to show that the probability that at least one event from $\mathcal{F}$ happens is *high*. Hence we can assume w.l.o.g. that $\mathcal{F}$ is the family of all our events.

Let us interpret $Event_i$ as subsets of the probabilistic space Ω. Then $p_i = |\, Event_i \,|/|\Omega|$. The following estimation of the cardinality of the sum of sets is a weak form of the so-called *inclusion–exclusion* principle:

$$|\bigcup_i Event_i \,| \geq \sum_i |\, Event_i \,| - \sum_{i \neq j} |\, Event_i \cap Event_j \,|.$$

We have

$$|\, Event_i \cap Event_j \,| = (|\, Event_i \,| \cdot |\, Event_j \,|)/|\Omega|$$

owing to the fact that the events are pairwise independent.

Hence the above form of the inclusion–exclusion principle implies that the probability that at least one event happens is at least

$$\sum_i p_i - \sum_{i \neq j} p_i p_j = \sum_i \Big(p_i \cdot \Big(1 - \sum_{j \neq i} p_j\Big)\Big).$$

Now, since $\sum p_i \leq 1/3$ we have that the probability that at least one event happens is at least

$$\left(\sum_i p_i\right) \cdot \left(1 - \frac{1}{3}\right) \geq \min\{c, 1/6\} \cdot \left(1 - \frac{1}{3}\right) = c'$$

□

4.6 Removing a large proportion of elements

Many randomized algorithms work by iteratively reducing a certain set of elements, and performing in the meantime some additional operations with respect to the reduced set. Assume that the operation

$$S := \text{Reduce}(S)$$

removes from S some number of elements. Such algorithms have the following general *algorithmic scheme*:

Algorithm Abstract_Scheme
$S :=$ a given set; {initially $|S| = n$}
while $S \neq \emptyset$ **do**
  begin
    $S :=$ Reduce(S);
    {other parts of the algorithm not affecting S}
  end

We will need later to know that this algorithm performs on average a logarithmic number of iterations.

We state the following useful theorem which is frequently used later.

Theorem 4.6.1
Assume that initially $|S| = n$*, and the expected number of elements removed from a set* S' *by operation Reduce*(S') *is bounded from below by* $\max\{\epsilon \cdot |S'|, 1\}$*, where* $\epsilon > 0$ *is a constant. Then the expected number of iterations in the algorithm* Abstract_Scheme *is* $O(\log n)$.

Proof Let X_i be the number of elements removed in one iteration and S_i the total number of elements deleted so far after stage i. We use the following probabilistic fact; for a technical proof we refer to e.g. [Ko92].

Claim Let $\epsilon > 0$, and $X_1, X_2, \ldots$ and $S_1, S_2, \ldots$ be sequences of random variables such that

$$S_i = \sum_{j=1}^{i} X_j \leq m \text{ and } E(X_{i+1} |\, S_i = s) \geq \epsilon \cdot (n - s).$$

Then the expected least k such that $S_k = m$ is $O(\log n)$.

The claim implies that the expected least k for which $S_k = n$ is $O(\log n)$. This implies that after stage $k = O(\log n)$ all elements are removed from the set S. □

4.7 Derandomization

Derandomization is a powerful tool that is used to construct deterministic NC-algorithms. Historically the first NC-algorithm for *maximal* independent sets was constructed in this way, and consequently the first NC-algorithm for *maximal* matching (as maximal independent sets in the edge graph).

If we have an RNC-algorithm $\mathcal{A}$ which uses only $k = O(\log n)$ *random bits* then it can be converted to a deterministic NC-algorithm $\mathcal{A}'$.

The algorithm $\mathcal{A}'$ simulates $\mathcal{A}$ for each possible sequence of k bits in parallel. In other words we have 2^k copies of $\mathcal{A}$ working in parallel, each one for a different combination of bits. These random bits are no longer random. They are fixed for each copy of $\mathcal{A}$. We can summarize this as follows.

Fact 4.7.1
Each RNC-algorithm using $k = O(\log n)$ random bits can be converted to an NC-algorithm by increasing the number of processors by a factor 2^k.

If an RNC-algorithm uses a linear number of bits, then it can happen that it needs the random events to be only *pairwise independent.* A typical example is Luby's algorithm [L86], see Chapter 10. A subtle point in the theorem is that the algorithm which we want to derandomize has a special property:

it works properly if the probabilities are *suitably* approximated.

The proof of the algorithm is an example of linear algebra at work.

Theorem 4.7.2
Assume an RNC-algorithm uses n random events $Event_1, \ldots, Event_n$ whose probabilities are $p_1, \ldots, p_n$. Assume that we can replace probabilities p_i by some approximate probabilities p'_i such that $|p'_i - p_i| \le 1/n$, for $1 \le i \le n$. Assume also that only pairwise independence of events E_i is needed. Then the algorithm can be converted to an equivalent NC-algorithm.

Proof From Fact 4.7.1 it is enough to reduce the number of random bits to logarithmic. Take any prime number p from the interval $[n+1 \ldots 2n]$. Define the probabilistic space

$$\Omega = \{(x, y) \ : \ 0 \le x, y < p\}.$$

The probability of each single point in Ω is the same. Let $\oplus$ be the addition *modulo* p. Define the function

$$f_i \ : \ \Omega \to [0..p-1], f_i(x, y) = x \oplus iy.$$

Approximate each p_i by the closest fraction of the form k_i/p. Denote $A_i = \{0 \ldots k_i\}$. Define the i-th event as

$$E_i = \{(x, y) : \ f_i(x, y) \in A_i\}.$$

We show that the probability of the i-th event is k_i/p and the events are pairwise independent.

Claim 1

$$|E_i|/|\Omega| = k_i/p.$$

The claim follows from the fact that for each $z \in [0..p-1]$ and $i \in A_i$ there is exactly one x which is a solution to

$$x \oplus iy = z.$$

Hence $|f_i^{-1}(A_i)| = p \cdot |A_i| = k \cdot k_i$.

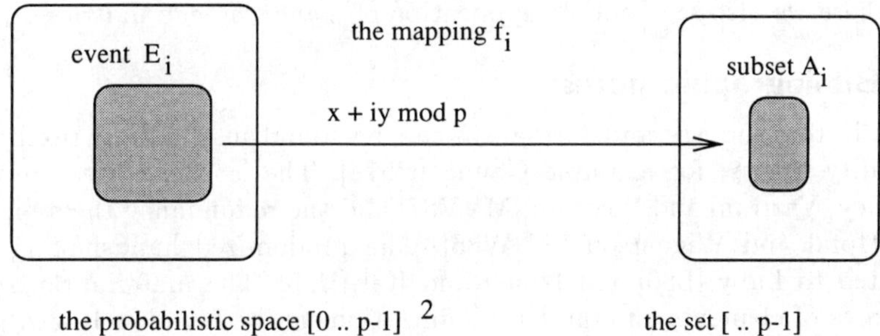

Fig. 4.3 The event E_i equals $\{(x, y) : f_i(x, y) \in A_i\}$, $|A_i \times A_j| = |E_i \cap E_j|$.

Claim 2
Let $i \neq j$. Then for each z_1, z_2 there is exactly one pair (x, y) which is a solution to

$$x \oplus iy = z_1$$
$$x \oplus iy = z_2.$$

This system of equations can be written in a matrix form as follows:

$$\begin{bmatrix} 1 & i \\ 1 & j \end{bmatrix} \times \begin{bmatrix} x \\ y \end{bmatrix} = \begin{bmatrix} z_1 \\ z_2 \end{bmatrix}.$$

The claim follows from the fact that the 2×2 matrix corresponding to the above system of linear equations is nonsingular in the field Z_p of numbers modulo p.

Claim 3
The events E_i, E_j are independent, if $i \neq j$:

$$|E_i \cap E_j|/|\Omega| = k_i/p \cdot k_j/p.$$

This claim follows from Claim 2, since we have $|A_i \times A_j| = k_i k_j$ integers z_1, z_2 in Claim 2. This equals the cardinality of $E_i \cap E_j$. Hence

$$|E_i \cap E_j|/\Omega = (k_i \cdot k_j)/p^2.$$

□

Remark The theorem can be extended to *d-wise independent* events, for any constant d. As probabilistic space we take $[0..p-1]^d$, and the i-th event is

$$E_i = \{(x_0, x_1, \ldots, x_{d-1}) \; : \; x_0 \oplus ix_1 \oplus i^2 x_2 \oplus \ldots \oplus i^{d-1} x_{d-1} \in A_i\}.$$

The d-wise independence is reduced to the nonsingularity of so called *Vandermonde* matrices. We introduce them in the next chapter (devoted more to algebra), where we also see another application of *Vandermonde* matrices.

4.8 Bibliographic notes

The basic terminology and formalism can be found in standard textbooks on probability theory, for example Chung [Ch74]. The isolating theorem is from Mulmuley, Vazirani and Vazirani [MVV87] and the redundancy theorem is from Karp, Upfal and Wigderson [KUW86]. The randomized handshaking can be attributed to Luby [L86] and Israeli and Itai [II86]. The material on removing large parts of elements and on derandomization can be found in Kozen [Ko92].

5

Algebraic tools

5.1 Determinants versus permanents

Assume $A = [a_{i,j}]_{1 \le i,j \le n}$ is an $n \times n$ integer matrix. For a permutation $\sigma = (i_1, \ldots, i_n)$ define $\text{product}_A(\sigma) = a_{1,i_1} \cdot \ldots \cdot a_{n,i_n}$. The *permanent* and *determinant* of A are defined as follows:

$$\text{perm}(A) = \sum\nolimits_\sigma \text{product}_A(\sigma),$$

$$\det(A) = \sum\nolimits_\sigma \text{sign}(\sigma) \cdot \text{product}_A(\sigma),$$

where the summations are over all permutations of $[1, \ldots, n]$.

The *sign* of the permutation σ, denoted by $\text{sign}(\sigma)$, equals -1 if the number of inversions in σ is odd, otherwise the sign equals $+1$. An *inversion* is a pair of position $i < j$ such that $\sigma(i) > \sigma(j)$. $\text{sign}(\sigma)$ also equals $(-1)^{\#\text{transp}(\sigma)}$, where $\#\text{transp}(\sigma)$ is the number of transpositions (of not necessarily adjacent elements) needed to obtain σ from the identity permutation $\text{id} = (1, 2, \ldots, n)$.

Lemma 5.1.1
The sign of an n-element permutation with exactly k even cycles is $(-1)^k$.

Recall that a bipartite graph $G = (V, W, E)$ is an undirected graph whose set of nodes consists of disjoint parts V, W, and E is a set of edges which connect some nodes of V to some nodes of W. If $|V| = |W|$ the the graph is called *equibipartite*.

Assume an $n \times n$ matrix A consists of zeros and ones then we define an equibipartite graph $G = \text{BiGraph}(A) = (V, W, E)$, where $V = \{v_1, \ldots, v_n\}$, $W = \{w_1, \ldots, w_n\}$ and $E = \{(v_i, w_j) : A[i,j] = 1\}$.

Then $\text{perm}(A)$ equals the number of perfect matchings of G. On the other hand $\det(A)$ can have no *sensible* relation to the number of perfect matchings.

Example Assume each entry of the matrix A equals 1.Then $\text{BiGraph}(A)$ is a complete bipartite graph and there are $n!$ perfect matchings. We have $\text{perm}(A) = n!$ but $\det(A) = 0$.

This would imply that the permanent is much more useful. However, computing the permanent seems to be an infeasible task, while determinants are easy to compute.

The class #P consists of very hard counting problems, for example like the computation of assignments of values in a Boolean formula such that the result is *true*. The problem of checking if this number is nonzero is NP-complete. The computation of permanents is in a certain sense at least as difficult as solving an NP-complete problem.

In fact it is easy to compute the determinant of the whole matrix together with determinants of all its $(n-1)\times(n-1)$ submatrices, called *minors*. The minor $A_{i,j}$ of A is a submatrix resulting by removing the i-th row and j-th column.

We assume that each arithmetic operation on the logarithmic bits numbers can be done in $O(1)$ time with one processor. However, if we deal with integers having a large number m of bits then the number of processors should be suitably multiplied.

Denote by $\Lambda(n)$ the number of processors needed to compute the determinant of an $n\times n$ matrix and its minors. It is known that $\Lambda(n)=O(n^{3.5}m)$ processors are sufficient if we deal with integers having $O(m\log(n))$ bits, see [P85].

We refer e.g. to [BGH82], [B84], [PS78], [GP89] and [Ko92] for the proof of the following lemma.

Lemma 5.1.2 ([PS78], [BGH82], [B84], [GP89], [P85])
(1) The problem of computing permanents of zero–one matrices is #P-complete.
(2) $\det(A)$ *together with determinants of all its minors can be computed in* $\log^2 n$ *parallel time with* $\Lambda(n)=O(n^{3.5}m)$ *processors if we deal with integers having* $O(m\log(n))$ *bits.*

The determinants are easier to compute since they have many useful algebraic properties. The determinant can be thought as a function of n vectors $[\bar{v}_1,\bar{v}_2,\ldots,\bar{v}_n]$, which are columns of A. The matrix A can be written as $A=[\bar{v}_1,\bar{v}_2,\ldots,\bar{v}_n]$. The function det can be characterized (in an equivalent way) as a function satisfying:

(1) $\det[\bar{v}_1,\ldots,a\bar{v}_i+b\bar{w}_i,\ldots\bar{v}_n]=$
$a\cdot\det[\bar{v}_1,\ldots,\bar{v}_i,\ldots\bar{v}_n]+b\cdot\det[\bar{v}_1,\ldots,\bar{w}_i,\ldots,\bar{v}_n]$;

(2) if $\bar{v}_i=\bar{v}_j$ for some $i\neq j$ then $\det[\bar{v}_1,\ldots,\bar{v}_n]=0$;

(3) $\det[\bar{j}_1,\ldots\bar{j}_n]=1$, where $\bar{j}_i$ is the i-th *column vector*, is the column vector whose i-th component is 1 and all other components are zero.

It follows that adding one row to the other, or adding one column to another (with any coefficient), does not change the value of the determinant. This gives a sequential method of computing the determinant which works like the Gauss elimination method for solving systems of linear equations. However, parallel algorithms work quite differently and are nontrivial.

We give some examples of determinants. A useful matrix is *Vandermonde's matrix* $\mathrm{Vand}(a_1,a_2,\ldots,a_t)$, which is of the form

$$\begin{bmatrix} a_1^0 & a_2^0 & a_3^0 & \dots & a_t^0 \\ a_1^1 & a_2^1 & a_3^1 & \dots & a_t^1 \\ a_1^2 & a_2^2 & a_3^2 & \dots & a_t^2 \\ \vdots & \vdots & \vdots & \vdots & \vdots \\ a_1^{t-1} & a_2^{t-1} & a_3^{t-1} & \dots & a_t^{t-1} \end{bmatrix}$$

We have (see [Co74])

$$\det(\mathrm{Vand}(a_1, a_2, \dots, a_t)) = \prod_{i<j}(a_j - a_i).$$

Two other examples of interesting determinants are presented below. Consider the following matrix:

$$A_n = \begin{bmatrix} 0 & 1 & 0 & 0 & 0 & 0 \\ -1 & 1 & 1 & 0 & 0 & 0 \\ 0 & -1 & 1 & 1 & 0 & 0 \\ 0 & 0 & -1 & 1 & 1 & 0 \\ \vdots & \vdots & \vdots & \vdots & \vdots & 1 \\ 0 & 0 & 0 & 0 & -1 & 1 \end{bmatrix}$$

The numbers $\det(A_n)$ are *Fibonacci numbers.* In the sequel we will frequently use matrices whose entries are variables and the values of determinants are (symbolic) polynomials. An example of such (symbolic) matrix is

$$A = \begin{bmatrix} x & 0 & 0 & \dots & 0 & a_0 \\ -1 & x & 0 & \dots & 0 & a_1 \\ 0 & -1 & x & \dots & 0 & a_2 \\ 0 & 0 & -1 & x & 0 & a_3 \\ \vdots & \vdots & \vdots & \vdots & x & a_{n-1} \\ 0 & 0 & 0 & \dots & -1 & a_n \end{bmatrix}$$

For this matrix A we have

$$\det(A) = a_0 + a_1x + a_2x^2 + a_3x^3 + \dots + a_nx^n.$$

5.2 Symbolic polynomials of set families

We consider large multivariate polynomials related to families $\mathcal{M}$ of subsets of a certain set E. Our main family $\mathcal{M}$ will be the family of maximum matchings and E will be the set of edges. The main idea is to associate with $\mathcal{M}$ some polynomial $\Psi(\mathcal{M})$ (called the symbolic polynomial of $\mathcal{M}$) which gives useful information about $\mathcal{M}$. Then using some algebraic techniques we can extract from $\Psi(\mathcal{M})$ information about $\mathcal{M}$. This resembles the concept of generating functions.

The basic-type polynomials $\mathcal{P} = \Psi(\mathcal{M})$ related to family $\mathcal{M}$ of maximum matchings in a graph are the so-called *Tutte's symbolic polynomials*, to be defined in Chapter 6. They correspond to determinants of matrices, whose elements are variables with integer coefficients. The monomials of considered polynomials are related directly to the family $\mathcal{M}$ of all perfect matchings of a given graph.

Partial interpolation means that we want to know at least one of the subsets of $\mathcal{M}$, and this corresponds to reconstructing one of the monomials of a polynomial (and corresponds to the construction of maximum matching).

In this section we introduce an abstraction of Tutte's polynomials: *special black-box polynomials.* The algebraic approach to parallel matching consists in using such polynomials. There is a whole theory about the interpolation of more general black-box polynomials. However, for special polynomials their relation to matchings is more apparent.

Each polynomial $\Psi(\mathcal{M})$ is a sum of terms, each term a coefficient (an integer) multiplied by a monomial (product of variables). The degree of a term is the sum of exponents of variables which appear in this term. The *degree* of the whole polynomial $\mathcal{P}$, written $\deg(\mathcal{P})$, is the maximum of the degrees of its terms.

We can associate with each element x of E a symbolic variable x; formally we can identify E with a set *Var* of variables. Then we can associate with a set a monomial by the following function mon:

$$\mathrm{mon}(\{x_1, x_2, \ldots, x_k\}) = x_1 x_2 \ldots x_k.$$

There are various ways to define transformation Ψ of set families into symbolic polynomials. The collection of sets can be, for example, transformed into a sum of monomials corresponding to members of $\mathcal{M}$, possibly with integer coefficients. Such a type of Ψ can be realized using Pfaffians (see Chapter 6). However, for technical reasons (to avoid computing square roots) we consider Ψ of the form

$$\Psi(\{M_1, M_2, \ldots, M_t\}) = \\ (a_1 \cdot \mathrm{mon}(M_1) + a_2 \cdot \mathrm{mon}(M_2) + \ldots + a_t \cdot \mathrm{mon}(M_t))^2, \qquad (5.1)$$

where $a_i = +1$ or $a_i = -1$. We introduce $\mathcal{M}$-consistent polynomials later to generalize symbolic polynomials of this form.

There are three types of information which we shall extract from $\mathcal{P} = \Psi(\mathcal{M})$:

1. Is the family $\mathcal{M}$ empty? This corresponds to a nonzero test for the polynomial $\mathcal{P}$.
2. Find any member $M_i \in \mathcal{M}$. This corresponds to finding one term of $\mathcal{P}$.
3. Find the maximum cardinality of a set in $\mathcal{M}$. This corresponds to the computation of $\deg(\mathcal{P})$.

We shall describe three algorithms related to the questions above:

(1) the *nonzero-test* algorithm;

(2) the *partial interpolation* algorithm;

(3) the *smallest degree* algorithm.

If we know a term of the form $\mathrm{mon}(M_i) \cdot \mathrm{mon}(M_i)$ then we can reconstruct the set M_i. This approach requires the arithmetic of exponential numbers.

The most efficient algorithm for general matchings uses the degree computation approach. The computation of $\deg(\mathcal{P})$ can be reduced to the computation of the smallest degree of certain one-variable polynomials related to $\mathcal{P}$. We use an algebraic trick: a new polynomial $\mathcal{P}'(z)$ is created by multiplying each term α in $\mathcal{P}$ by $z^{k-\deg(\alpha)}$, where z is a new variable and k is sufficiently large. Then

$$\deg(\mathcal{P}) = k - \text{smallest\_degree}_z(\mathcal{P}'(z)).$$

In this chapter we consider only the computation of the smallest degree; the construction of $\mathcal{P}'(z)$ is considered in the next chapter.

The polynomials related to families $\mathcal{M}$ (with potentially exponentially many members) cannot be explicitly manipulated. We only have partial information about them, the main information being the values of such polynomials.

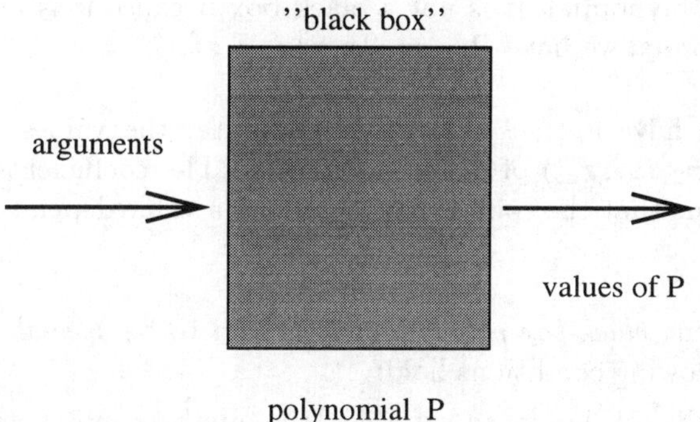

Fig. 5.1 A *black box* contains an unknown polynomial $\mathcal{P}$: we do not know $\mathcal{P}$ but we can compute $\mathcal{P}(\bar{x})$ efficiently.

Assume we have a *black box* which gives the values of a polynomial; we call it a *black-box polynomial.* The *black box* means that we do not have an explicit representation of the polynomial; we do not even know whether this polynomial is nonzero ("nonempty"). However, we can evaluate this polynomial at any given point. This corresponds to questions to the black box of the type "*what is the value of* $\mathcal{P}(\bar{x}_0)$?" The black box is obliged to answer correctly. Using such answers we want to know what is in the black box. This is also related to learning theory: learning unknown object by queries.

Reconstructing the whole polynomial is called *polynomial interpolation.* Usually we do need the *whole truth*; sometimes it is enough to know if the black box is *empty* or to know at least a small piece of it, that is, one monomial, called *partial interpolation.* In this section we show how to test if a black box is nonzero.

A polynomial $\mathcal{P}(\bar{x})$ is said to be $\mathcal{M}$-*consistent* iff all its monomials are some elements of $\{\text{mon}(X)\cdot\text{mon}(Y) \;:\; X, Y \in \mathcal{M}\}$ with integer coefficients. Moreover, we require that for each $X \in \mathcal{M}$ the monomial $\text{mon}(X)^2$ appears in $\mathcal{P}$ and its coefficient in $\mathcal{P}$ has absolute value equal to 1.

Hence an *$\mathcal{M}$-consistent polynomial* is of the form

$$P(\bar{x}) = \sum_{M_i, M_j \in \mathcal{M}} a_{i,j} \cdot \text{mon}(M_i) \cdot \text{mon}(M_j),$$

where $|a_{i,i}| = 1$ for each $M_i \in \mathcal{M}$ and all $a_{i,j}$ are integers.

For a given variable x denote by $\mathcal{P}_x$ the polynomial consisting of all monomials of $\mathcal{P}$ containing the variable x. For a subset $Z \subseteq Var$ denote by $\mathcal{P}_Z$ the subpolynomial of $\mathcal{P}$ consisting of all monomials containing all variables from the set Z.

Example Let $Var = \{x_1, x_2, x_3\}$ and $\mathcal{M} = \{\{x_1, x_2\}, \{x_2, x_3\}\}$. Then the polynomial $\mathcal{P}(x_1, x_2, x_3) = x_1^2 x_2^2 + 2x_1 x_2^2 x_3 - x_2^2 x_3^2$ is an example of a special $\mathcal{M}$-consistent polynomial. It is not a black box because it is *explicitly* written. For this polynomial we have $\mathcal{P}_{x_3} = 2x_1 x_2^2 x_3 - x_2^2 x_3^2$.

Assume we have a *black box* which computes the values of a polynomial $\mathcal{P}(\bar{x}) = \mathcal{P}(x_1, x_2, \ldots, x_m)$ of degree at most n. The coefficients and values are integers. Assume that the complexity of getting a value depends on a parameter n.

An $\mathcal{M}$-consistent *black-box polynomial* $\mathcal{P}$ is said to be *special* with complexity $P(n)$ if the following conditions hold:

(a) we know the set Var of m variables, the numbers m depends on n;

(b) we know that $\mathcal{P}$ is $\mathcal{M}$-consistent for some (unknown) family $\mathcal{M}$ of subsets of Var;

(c) for each value $\bar{x}_0$ we can compute values $\mathcal{P}(\bar{x}_0)$ and $\mathcal{P}_x(\bar{x}_0)$, for each $x \in Var$, in $O(\log^2 n)$ time with $O(P(n))$ processors;

(d) for each $X \subseteq Var$ we can test if $X \in \mathcal{M}$ in $O(\log n)$ time with $O(P(n))$ processors.

The basic information which we shall need about the black box $\mathcal{P}$ is any element of the family $\mathcal{M}$. Such an element will be seen later to correspond to a perfect matching (see Chapter 6).

5.3 Nonzero test and partial interpolation of polynomials

In this section we show how to compute some partial information about a special black-box polynomial. We perform a partial reconstruction of a polynomial knowing how to compute its values; usually it is related to the polynomial interpolation. There are two types of partial information which we need:

1. The weakest one is to answer if $\mathcal{P}$ is not identically zero (*nonzero test*).
2. The other one is to find any monomial of $\mathcal{P}$ (called *partial interpolation*); it is related to finding a set in the family $\mathcal{M}$.

Denote by *Var* the set of variables of a polynomial $\mathcal{P}$. The following basic theorem is a version of theorems given by Schwartz and (independently) by Zippel, see [S80] and [Z79]. There is one subtle point about this theorem.

The polynomial $\mathcal{P}$ can be nonzero but it can have infinitely many zeros (roots) (this cannot happen to one-variable polynomials). The reason is very simple: one of the variables can be absent in $\mathcal{P}$. Then if $\mathcal{P}$ has a root we can change the value of this variable arbitrarily. The next theorem says essentially that the number of roots is anyway *small.*

Theorem 5.3.1 (nonzero-test theorem)
Let $\mathcal{P}$ be a nonzero black-box polynomial of degree at most n with m variables. Assume that we assign to each variable in Var a random value from a set Ω of integers of cardinality N. Then

$$Prob\{\mathcal{P}(\bar{x}) \neq 0\} \geq 1 - \frac{n}{N}.$$

Proof From the equality

$$\frac{n \cdot N^{m-1}}{|\Omega|^m} = \frac{n}{N}$$

it is enough to prove the following.

Claim $\mathcal{P}$ has at most $n \cdot N^{m-1}$ roots in the set Ω^m.

The proof of the claim is by induction on m, the number of variables. The case $m = 1$ is trivial so assume $m > 1$ and the claim is valid for any $m' < m$. Assume (without loss of generality) that the variable x_1 appears in $\mathcal{P}$. Then we can write $\mathcal{P}$ in the form

$$\mathcal{P}(x_1, x_2, \ldots, x_m) = \sum_{i=0}^{k} x_1^i \mathcal{Q}_i(x_2, \ldots, x_m),$$

where k is the maximal degree of x_1 in any term of $\mathcal{P}$ and $\mathcal{Q}_k$ is a nonzero polynomial of $m-1$ variables of degree at most $n-k$.

Owning to the equality $(n-k) \cdot N^{m-1} + k \cdot N^{m-1} = n \cdot N^{m-1}$ the thesis follows from the following two facts:

(1) there are at most $(n-k) \cdot N^{m-1}$ roots of $\mathcal{P}$ which are also roots of $\mathcal{Q}_k$;
(2) there are at most $k \cdot N^{m-1}$ roots of $\mathcal{P}$ which are not roots of $\mathcal{Q}_k$.

The first point follows by the inductive assumption since $\mathcal{Q}_k$ is a polynomial of degree at most $n-k$ and having $m-1$ variables. It is bounded by $(n-k)N^{m-2} \leq (n-k)N^{m-1}$.

The second point follows from the fact that for each assignment of values to variables $(x_2, \ldots, x_m)$, if $\mathcal{Q}_k$ does not disappear then we have a nonzero univariate polynomial of degree k with respect to x_1. Such a polynomial has at most k distinct roots.

□

Theorem 5.3.2
Assume P is a special black-box polynomial with complexity $P(n)$ and degree at most n. Then we can perform the nonzero test for P by a randomized algorithm working in $\log^2(n)$ time with $P(n)$ processors.

Proof This is a consequence of Theorem 5.3.1.

□

The basic information which we shall need about the black box $\mathcal{P}$ is any element of the family $\mathcal{M}$.

The following theorem was essentially shown in [MVV87], though the terminology was different there.

Theorem 5.3.3 (partial-interpolation theorem)
If a polynomial black-box $\mathcal{P}$ is special with complexity $P(n)$ then we can reconstruct one of the monomials of $\mathcal{P}$ in randomized time $O(\log^2 n)$ with $O(P(n))$ processors.

Proof We use the *isolating theorem* for the family $\mathcal{M}$ and the terminology used in this lemma. The powers of two are introduced to translate summation (in the isolating theorem) into multiplication (in monomials of the polynomial $\mathcal{P}$).

A random number $\hat{x}$ from the set $[1 \ldots 2|\,Var\,|]$ is picked for each variable $x \in Var$. The value $2^{\hat{x}}$ is assigned to each variable x. A vector $\bar{x}_0$ of values assigned to variables is generated in this way and the polynomial $\mathcal{P}$ can be computed at point $\bar{x}_0$.

Using these ideas we construct the algorithm *Partial_Interpolation* described below. There is a large probability that after one iteration of this algorithm a correct answer is returned.

A random number $\hat{x}$ from the set $[1 \ldots 2|\,Var\,|]$ picked for the variable $x \in Var$ is the *weight* of x. Each subset Z of variables has its weight $\hat{Z}$ as the sum of weights of its elements. By the isolating theorem there is a large probability that the subset X with minimum weight r is unique. Observe also that the coefficient at $\text{mon}(X, X)$ is 1. Then the weight of all other subsets in $\mathcal{M}$ is bigger than the weight of X. Hence the highest power r of two which divides the value of the whole polynomial is the power corresponding to the "smallest" monomial $\text{mon}(X, X)$. This value r is recovered in step 2 of the algorithm.

Hence the values of all monomials other than $\text{mon}(X, X)$ are larger powers of two. Now variable x is in this *smallest* monomial iff $\mathcal{P}_x(\bar{x}_0)$ is a sum of powers of two such that the lowest of them is the value 2^r. This is tested in step 2. From the isolating theorem we succeed with high probability in one iteration.

□

```
Algorithm Partial_Interpolation;
  Step 1.
     pick a random number x̂ from [1...2m] for each x ∈ Var;
     assign to each x the value 2^x̂;
     {in this way a vector x̄0 of values assigned to variables is created}
     compute val = P(x̄0);
     for each x ∈ Var do in parallel
        compute val_x := P_x(x̄0);
  Step 2.
     let r be the largest integer such that 2^r divides val;
     X:= {x : val_x mod 2^r = 1};
  Step 3.
     if X ∈ M then return X and STOP;
```

The following lemma is an extension of the isolating theorem, the proof (almost the same as in the original lemma) is omitted.

Lemma 5.3.4 (weighted isolating lemma)
Assume we have a family $\mathcal{M}$ of k-subsets of the set $E = [1 \ldots m]$, each element x of E having a nonnegative integer weight $w(x)$ assigned to it. We assign to each x a new weight $2mk \cdot w(x) + random(x)$, where $random(x)$ is a random element of $[1 \ldots 2m]$. Then with probability at least $1/2$ there is unique subset in $\mathcal{M}$ of minimal weight with respect to the new weight. This subset is also minimal with respect to the original weight.

Assume we assigned integer weights to variables. The weight of a monomial τ is the sum of weights of variables contained in τ. If the variable appers twice then its weight is counted twice. The constant coefficient is diregarded and treated as 1. Using the weighted isolating theorem we can easily extend the partial-interpolation theorem to the weighted case as follows:

Theorem 5.3.5 (weighted partial-interpolation theorem)
Assume that to each variable $x \in Var$ there is assigned an integer weight(x) given in unary. If a polynomial black box $\mathcal{P}$ is special with complexity $P(n)$ then we can reconstruct a smallest-weight polynomial $\mathcal{P}$ in randomized time $O(\log^2 n)$ with $O(P(n))$ processors.

5.4 Sparse polynomials

In this section we show that the nonzero-test for polynomially-sparse polynomials can be done by a deterministic NC-algorithm. In fact such polynomials can be totally reconstructed by an NC-algorithm. We shall see in the next chapter how this is done for some matrices corresponding to graphs.

We say that a polynomial black box $\mathcal{P}$ is *polynomially sparse* iff the number of its nonzero monomials is polynomially bounded. This translates into the statement: $\mathcal{P}$ is $\mathcal{M}$-consistent, where $|\mathcal{M}|$ is bounded by a polynomial.

We need the following algebraic lemma based on properties of Vandermonde matrices.

Lemma 5.4.1 (nonzero-sum lemma)
Assume the numbers $\alpha_1, \alpha_2, \ldots, \alpha_t$ are pairwise distinct and $(a_1, a_2, \ldots, a_t)$ is a nonzero vector. Then there exists $0 \le s < t$ such that:

$$sum_s \;=\; \sum_{i=1}^{t} a_i \alpha_i^s \;\neq\; 0.$$

Proof Observe that

$$\begin{bmatrix} \alpha_1^0 & \alpha_2^0 & \alpha_3^0 & \ldots & \alpha_t^0 \\ \alpha_1^1 & \alpha_2^1 & \alpha_3^1 & \ldots & \alpha_t^1 \\ \alpha_1^2 & \alpha_2^0 & \alpha_2^0 & \ldots & \alpha_t^2 \\ \vdots & \vdots & \vdots & \ddots & \vdots \\ \alpha_1^{t-1} & \alpha_2^{t-1} & \alpha_3^{t-1} & \ldots & \alpha_t^{t-1} \end{bmatrix} \times \begin{bmatrix} a_1 \\ a_2 \\ a_3 \\ \vdots \\ a_t \end{bmatrix} =$$

$$[\ \mathrm{sum}_0,\ \ \mathrm{sum}_1,\ \ \mathrm{sum}_2,\ \ \ldots,\ \ \mathrm{sum}_{t-1}\ ]$$

The matrix on the left is a *Vandermonde matrix*; it is nonsingular (has nonzero determinant) for distinct values $\alpha_1, \alpha_2, \ldots, \alpha_t$. It is known that multiplication of a square nonsingular matrix by a nonzero vector gives a nonzero vector. Since $(a_1, a_2, \ldots, a_t)$ is nonzero we obtain that the resulting vector is nonzero. Hence at least one of the numbers $\mathrm{sum}_0, \ldots, \mathrm{sum}_{t-1}$ is nonzero.

□

Observation The value $t-1$ is generally the minimal possible value in Lemma 5.4.1. There are pairwise distinct numbers $\alpha_1, \alpha_2, \ldots, \alpha_t$ and a nonzero vector $(a_1, a_2, \ldots, a_t)$ such that $\sum_{i=1}^{t} a_i \alpha_i^{t-1} \neq 0$, but for each $0 \le s < t-1$

$$\mathrm{sum}_s \;=\; \sum_{i=1}^{t} a_i \alpha_i^s \;=\; 0.$$

We can fix

$$(\mathrm{sum}_0, \ldots, \mathrm{sum}_{t-1}) = (0, 0, \ldots, 0, 1)$$

and find the corresponding nonzero vector $(a_1, a_2, \ldots, a_t)$ by solving the system of linear equations. This is possible since the matrix is nonsingular.

The nonzero-sum lemma implies that we can test $\mathcal{P} = 0$ for polynomially sparse polynomials by an NC-algorithm. In fact we can do much more than that (reconstructing the whole sparse polynomial) using deterministic NC-computations, but it is beyond the scope of this text and it is not needed later.

Theorem 5.4.2 (sparse-test theorem, [GK87])
Assume we know that an $\mathcal{M}$-consistent black-box polynomial $\mathcal{P}$ is polynomially sparse: the number of different terms of $\mathcal{P}$ is bounded by n^k, where k is known. Then we can test, by an NC-algorithm, whether $\mathcal{P}$ is a nonzero polynomial.

Proof Assign to each variable x_i a distinct prime number p_i. Then our polynomial can be written as

$$P(x_1, \ldots, x_m) = \sum_{i=1}^{t} a_i \cdot \mathrm{mon}_i(x_1, \ldots, x_m)$$

where $t = n^k$ and mon_i are monomials. After substituting prime numbers for the x_i the value of $\mathrm{mon}_i(x_1, \ldots, x_m)$ becomes α_i. Because we substitute different prime numbers all the α_i are pairwise different. Now we can take $t = n^k$ in Lemma 5.4.1, which implies the following:

Claim If the polynomial $\mathcal{P}$ is nonzero then, for $0 \leq s < n^k$, at least one of the values $\mathcal{P}(p_1^s, p_2^s, \ldots, p_m^s)$ is nonzero.

Hence it is enough to compute the polynomial at t different points to check if it is nonzero. Such computations can be done independently in parallel.

□

5.5 Randomized computation of the smallest degree

Assume $\mathcal{P} = \mathcal{P}(x_1, x_2, \ldots, x_m) \neq 0$ is a black-box polynomial, and z is one of the variables x_i. Denote by $\mathrm{mindeg}_z(\mathcal{P})$ the minimal degree of of a nonzero term of $\mathcal{P}$ with respect to a variable z. This means that if other variables are treated as coefficients, $\mathcal{P}$ becomes a one-variable polynomial $\mathcal{P}(z)$, and $\mathrm{mindeg}_z(\mathcal{P})$ is the minimal degree of a term of such a polynomial.

The practical meaning of parameters r and c (in the lemma below) will be seen later.

Lemma 5.5.1 (number-theoretic lemma)
Assume $r > 4$ is a constant. Let p be a random prime number from the interval $[n \ldots n^r]$. Let a be a positive integer such that $a \leq n^{(r+c)n}$ for a constant c. Then, for a constant c' we have

$$Prob\{p \text{ divides } a\} \leq c' \cdot \log(n)/n^{r-1}.$$

Proof The number a has at most $\alpha = (r + c)n$ different prime factors larger than n. On the other hand it is known that there are at least $\beta = c''n^r / \log_2(n^r)$ prime numbers in the interval $[n \ldots n^r]$, see [MR95].

Hence the probability that a randomly chosen prime appears in the factorization of a is at most α/β, which gives, after simple calculations, the required estimation.

□

Example Take $n = 10$ and $r = 5$; then the number a has at most $rn = 50$ possible distinct prime factors, and though a can be astronomically large, it can be close to 10^5. There are 9588 primes in the interval $[10 \ldots 10^r]$.

Hence if $a \in [1 \ldots 10^{50}]$ then

$$Prob\{\text{a random prime number from } [10 \ldots 10^5] \text{ divides } a\} \leq \frac{50}{9588} < 0.006.$$

We relate the degree $\text{mindeg}_x(\mathcal{P})$ to powers of some prime numbers. For a prime number p and an integer x define the *p-adic order* of x as

$$[x]_p = \max\{j \ : \ p^j \text{ divides } x\}.$$

Let $\mathcal{Q}(z) \neq 0$ be a one-variable polynomial with integer coefficients. The p-adic order of $\mathcal{Q}(z)$ is defined as the least p-adic order of coefficients of $\mathcal{Q}(z)$, and is denoted by $[\mathcal{Q}(z)]_p$. We relate the smallest degree to p-adic order using the number-theoretic result presented earlier.

Lemma 5.5.2 (smallest degree of one-variable polynomials)
Assume $r > 4$ and $c > 0$ are constants. Let $\mathcal{Q}(z) = \sum_{i=j}^{n} a_i z^i$ be a one-variable polynomial with integer coefficients $a_i \leq n^{(r+c)n}$, for $j \leq i \leq n$ and with $a_j \neq 0$. Let p be a random prime number from the interval $[n \ldots n^r]$. Then, for a constant c' we have

$$Prob\{[\mathcal{Q}(p)]_p \neq mindeg_z(P)\} \leq c' \cdot \log(n)/n^{r-1}.$$

Proof Observe that

$$[\mathcal{Q}(p)]_p \neq \text{mindeg}_z(P) \Leftrightarrow p \text{ divides } a_j.$$

However, for a random prime number from the interval $[n \ldots n^r]$ p divides a_j with probability at most $c' \cdot \log(n)/n^{r-1}$ for a constant c', according to the Lemma 5.5.1.

□

Algorithm Smallest-Degree;

Step 1.
assign to each variable $x \neq z$ a random value from $[1 \ldots n^r]$;

Step 2.
assign to z a random prime number $p \in [n \ldots n^r]$;
{a vector $\bar{x}_0$ of values assigned to variables has been created}

Step 3.
compute *value* $:= \mathcal{P}(\bar{x}_0)$;

Step 4.
$d := [value]_p$;
return $\text{mindeg}_z(\mathcal{P}) = d$

Theorem 5.5.3 (degree computation theorem, [GP88])
Assume that $\mathcal{P}(x)$ is a special black-box polynomial with complexity $P(n)$. Then $mindeg_z(\mathcal{P})$ can be computed in randomized $log^2 n$ time with $O(P(n))$ processors.

Proof Take a constant $r > 4$. The algorithm *Smallest-Degree* computing the smallest degree is described above. There exists a representation of the polynomial $\mathcal{P}$ in the form

$$\mathcal{P}(\bar{x}) = \sum_{i=j}^{n} \mathcal{Q}_i(\bar{z}) \cdot z^i,$$

where $\bar{z}$ is a vector of all variables except z, and $\mathcal{Q}_i(\bar{z})$ are polynomials of $\bar{z}$ such that $j = \text{mindeg}_z(P)$ and $\mathcal{Q}_j$ is a nonzero polynomial of degree at most n.

The probability that after Step 1 the value $a = \mathcal{Q}_j(\bar{z}_0)$ is nonzero is at least $1 - n^{r-1}$. In this moment, before assigning a value to variable z, we have a polynomial $\mathcal{Q}$ of z, whose minimal degree is j. The value $\text{mindeg}_z(\mathcal{Q})$ is now computed with overwhelming probability according to the lemma about the smallest degrees of one-variable polynomials. The p-adic order of a given number can be computed in $log^2 n$ time with a linear number of processors.

The algorithm gives a correct answer in one iteration with very high probability.

□

5.6 Ranks of matrices

Another important concept related to determinants and used in the next chapter is the *rank* of the matrix. Let A be an $n \times n$ matrix. The matrix is *nonsingular* if its determinant is nonzero.

First we extend the notion of a minor, defined in the previous section. Let α, β be subsets of $[1 \ldots n]$ of the same cardinality k. Denote by $A_{\alpha,\beta}$ the submatrix which results from A by taking all entries which are simultaneously in a row whose index is in α and in a column whose index is in β. The number k is the *dimension* of $A_{\alpha,\beta}$. In other words

$$A_{\alpha,\beta} = [a_{i,j}]_{i \in \alpha, j \in \beta}.$$

Square submatrices of A of the form above are called *minors* of A. The *rank* of A, denoted by $\text{rank}(A)$, is the maximal dimension of its nonsingular minor. In other words

$$\text{rank}(A) = \max\{k \;:\; k = |\alpha| = |\beta| \text{ and } \det(A_{\alpha,\beta}) \neq 0, \text{ for some } \alpha, \beta \subseteq [1 \ldots n]\}.$$

In the definition above the sets α, β could be different. There is one important case when we can assume that these sets are equal. A matrix A is called *skew-symmetric* iff $a_{i,j} = -a_{j,i}$ for each $1 \leq i, j \leq n$. The basic matrices introduced in the next chapter, Tutte matrices, are skew-symmetric.

The relation of submatrices to matchings is clearly visible if we deal with permanents (instead of determinants) and equibipartite graphs. The following lemma follows trivially from definitions:

Lemma 5.6.1
Assume A is the adjacency matrix of a bipartite graph $G = (V, W, E)$: $a_{i,j} = 1$ iff $(v_i, w_j) \in E$. Then the size of a maximum cardinality matching in G equals the maximal dimension of a minor of A, whose permanent is nonzero. In other words it equals

$$\max\{k \;:\; k = |\alpha| = |\beta| \textit{ and } perm(A_{\alpha,\beta}) \neq 0, \textit{ for some } \alpha, \beta \subseteq [1 \ldots n]\}.$$

Unfortunately the computation of permanents is hard, and we have to deal with determinants. However, the nonzero test for permanents will be replaced by a randomized nonzero test for determinants and the number k from the lemma can be reduced to randomized rank computation.

Lemma 5.6.2 (skew symmetry lemma)
If A is skew-symmetric then

$$\text{rank}(A) = \max\{k \;:\; k = |\alpha| \textit{ and } \det(A_{\alpha,\alpha}) \neq 0, \textit{ for some } \alpha \subseteq [1 \ldots n]\}.$$

Proof The proof relies on an old theorem of Frobenius. The details are omitted here.

□

We refer the reader to [Ko92] for the proof of the following theorem (stated below) in this section. The proof involves some concepts of linear algebra, which we do not introduce in this text (as they are not needed in other parts): namely,the *characteristic polynomial* of the matrix, *eigenvalues*, *eigenspaces* and symmetric matrices. The proof would include a good part of a linear algebra course and is omitted here.

Theorem 5.6.3 (the rank-computation theorem)
The rank of an integer matrix can be computed by an NC-algorithm.

5.7 NC-computation of determinants

In this section we show (for completeness) that there exists an NC-algorithm computing the *determinant*. The algorithm is less efficient than previously cited algorithms, but deserves description due to its simplicity, while the exposition of the best known (usually quite complicated) algorithms is beyond the scope of this book. Our exposoition is a version of the construction presented in Mahajan and Vinay, A combinatorial algorithm for determinants, *SODA97*, pp. 730–738. The determinant computation is reduced to path computations in directed acyclic graphs (dag's, in short), which can be easily done in parallel. Assume we have a dag G with integer weights associated to its edges. The weight of a path is a product of weights associated with the edges of this path, in the order corresponding to the path. The *path computation* problem can be stated as follows:

> given a dag G and two nodes s and t, compute the sum of all weights of directed paths from s to t.

Lemma 5.7.1 (path computation)
There is an NC-algorithm for the path computation problem.

Proof We can assume w.l.o.g. that G consists of a number (which is a power of two) of layers. At the k-th iteration we compute the matrix corresponding to sums of path weights between each two nodes contained in layers which are at distance 2^k. The matrix for the next iteration is produced by (essentially) squaring the matrix from the previous iteration. In this way the problem is reduced to matrix multiplication.

□

We can interpret the determinant computation problem as computing the total weight of all directed cycle covers of a directed graph G. If we are given an $n \times n$ matrix A, then the corresponding graph is a complete directed graph with nodes $\{1, \ldots, n\}$, and the weights of the edges are defined as

$$\text{weight}(i, j) \;=\; A[i, j].$$

Assume for simplicity that n is even. Denote by $l(\mathcal{C})$ the number of all components of a directed cycle cover $\mathcal{C}$ and by *product*$(\mathcal{C})$ the weight (product of edge weights) of $\mathcal{C}$. The sign of the cycle cover is defined as $(-1)^{l(\mathcal{C})}$. Then it is easy to see the following:

$$\det(A) \;=\; \sum_{\mathcal{C} \text{ is a directed cycle cover}} \text{sign}(\mathcal{C}) \cdot \text{product}(\mathcal{C})$$

We create a directed acyclic graph G' corresponding to G; its nodes are tuple (u, i, v) plus two additional nodes s (start) and t (termination). The path in G' corresponds to a traversal of consecutive cycles of a directed cycle cover of G together with the special edges "jumping" from one cycle to another; the node when we start on a given cycle is called the *head* of the cycle. We require that the head of the cycle is the smallest node in the cycle.

The paths are constructed in such a way that the heads of consecutive cycles are in increasing order (other nodes can be in any order).

The graph G' can be better viewed as a *finite automaton*, with initial state s, terminal state t. The node (u, i, v) of G' (state of an automaton) keeps (in its "memory") the following information:

- u is the currently visited node on a cycle whose traversal started at v;
- i is the number of traversed original edges so far;
- v is the head of current cycle.

Figure 5.2 shows how the weight of a directed cycle cover corresponds to the weight of a path in the auxiliary graph.

Observation The number of special edges ("jump" edges) weighted -1 has the same parity as the sign of the permutation corresponding to the cycle cover, if n is even. It is the same as the parity of the number of segments, see Figure 5.2.

Formally the graph $G' \;=\; (V', E')$ for $G \;=\; (V, E)$ is defined as follows.

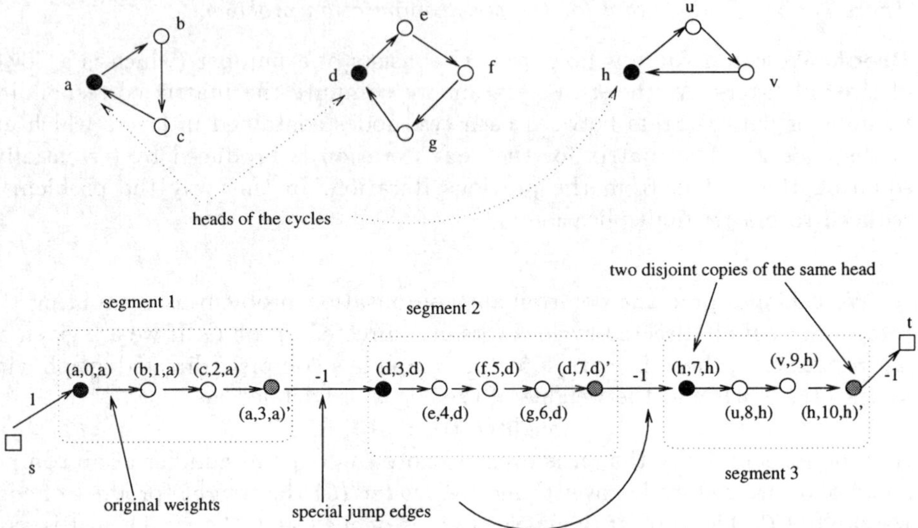

Fig. 5.2 An example of a directed cycle cover and a path in G' corresponding to this cover. Observe that end nodes in each segment are *primed*, so each head of a cycle has its two copies in the corresponding segment (starting copy and end copy). This is to ensure that there are exactly two occurrences of the head.

$$V' = \{(u,i,v) \; : \; u,v \in V, \; 0 \le i \le n\} \cup \{s, \; t\} \cup \{(v,i,v)' \; : \; v \in V, \; 0 \le i \le n\}.$$

The nodes of type (v,i,v) are beginnings of cycles, and $(v,i,v)'$ are end nodes of cycles.

There are 4 types of edges of G'

- **original edges:** $(u,i,v) \to (u',i+1,v)$, where $(u,u') \in E$, and $u' > v$, the weight of such edge equals weight(u,u');
- **original edges finishing cycles:** $(u,i,v) \to (v,i+1,v)'$, where $(u,v) \in E$, the weight of such edge equals weight(u,v);
- **start edges:** $s \to (u,0,u)$ of weight 1;
- **jump edges:** $(u,i,u)' \to (v,i,v)$, where $i > 0$ and $u < v$, we always "jump" from a smaller head (its primed copy) to a larger one (its starting copy);
- **end edges:** $(u,n,u)' \to t$, of weight -1;

Lemma 5.7.2
The sum of path weights from s to t in G' equals the sum of directed cycle cover weights of G multiplied by the signs of the corresponding cycle covers.

Proof Obviously each directed cycle cover C of G corresponds directly to a path π of G' such that product(C) $=$ product(π).

However there are paths π from s to t in G' which correspond to no correct directed cycle cover of G. Such paths are called here *bad paths.*

Claim For each sequence π from s to t in G' we can define a sequence $\Phi(\pi)$in such a way that the function Φ satisfies the following conditions.

1. π is a bad path if and only if $\Phi(\pi) \neq \pi$,
2. if π is a bad path then $\text{product}(\pi) = -\,\text{product}(\Phi(\pi))$,
3. $\Phi(\Phi(\pi)) = \pi$.

Proof (of the claim)
We partition the path π into maximal segements of consecutive *original* edges (disregarding jump edges)

$$\pi_1, \pi_2, \ldots, \pi_k,$$

see Figure 5.2. Now we can consider only the first components of nodes in segments. The sequence $\pi_1, \pi_2, \ldots, \pi_k$ corresponds now to the sequence of closed walks (clow's, in short), which start at some head and terminate at the head. In the example in Figure 5.2 we have the clow sequence
(a,b,c,a), (d,e,f,g,d), (h,u,v,h), where
head(a,b,c,a)=a, head(d,e,f,g,d)=d and head(h,u,v,h)=h.

The head of the clow is the smallest node: it appears only as the starting and terminating node of a clow. All other nodes in the clow are larger than the head. The heads of consecutive clows are strictly increasing. We call such sequences *correct clow seqences.* Each path in G' from s to t can be identifid with a correct clow sequence, and vice versa.

So let π be identified with its clow sequence $\pi_1, \pi_2, \ldots, \pi_k$.
If π is a directed cycle cover then we define $\Phi(\pi) = \pi$.
Otherwise find the largest i such that $\pi_i, \pi_{i+1}, \ldots, \pi_k$ is not a collection of disjoint simple cycles.

Go along π_i, starting at its head, until we find *the first* node w which violates the property of a good path. There are two cases.
Case 1: w appears in a clow π_j with $j > i$. Then we remove π_j from the clow sequence and insert it into π_i starting at w, suitably shifted, so that it ends in π_i and starts with w. The new clow sequence corresponds to a new path $\Phi(\pi)$. The parity of its components has changed w.r.t. π.
Case 2: w is not a head and terminates a simple subcycle $(w, u_1, u_2, \ldots, w)$. We remove this cycle from π_i, and create a new clow (which is a good simple cycle). We insert it into the correct place in the current clow sequence, according to its head (which is the smallest node in $(w, u_1, u_2, \ldots, w)$. The new clow sequence corresponds to the path $\Phi(\pi)$.
In this way we have described the transformation Φ. It is easy to see that it agrees with the conditions imposed by the claim. This completes the proof of the claim.

□

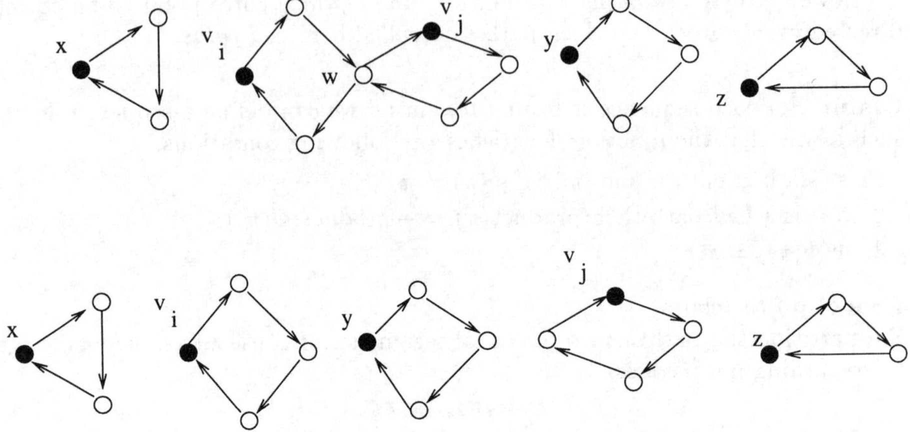

Fig. 5.3 Two clow sequences related through the function Φ. Assume that the heads of clow's are ordered $x < v_i < y < v_j < z$.

It follows directly from the claim that the total sum of products on all bad paths equals zero. The contribution of a bad path π is cancelled by the contribution of $\Phi(\pi)$, and vice versa, since the weights of these paths differ only in the sign (their number of segements differs by one). Now the correctness of the whole transformation follows from this fact. This completes the proof. □

Lemma 5.7.2 and Lemma 5.7.1 imply directly the following result.

Theorem 5.7.3
There is an NC-algorithm for computing determinants.

5.8 Bibliographic notes

Fast parallel algorithms for the computation of determinants can be found in Preparata and Sarwate [PS78], Borodin, von zur Gathen and Hopcroft [BGH82], Galil and Pan [GP89], Pan [P85]. The partial interpolation (under a different name) is developed in Mulmuley, Vazirani and Vazirani [MVV87]. The concept of sparse interpolation was introduced in Grigoriev and Karpinski [GK87]. The smallest-degree algorithm was introduced in Galil and Pan [GP88]. The material on the parallel computation of ranks of matrices is nicely presented in Kozen [Ko92], see also Berkowitz [B84]. A new algorithm for computing determinants was discovered recently in Mahajan and Vinay [MV97]. An interested reader in approximating permanents is referred to Jerrum and Sinclair [JS89].

6

Maximum cardinality matchings

In this chapter we use algebraic tools (in particular parallel interpolation algorithms for large polynomials) to find maximum cardinality matchings in undirected graphs. We explore the correspondence between graphs and some black-box polynomials (Tutte's symbolic polynomials).

The crucial point is the use of determinants instead of permanents. There are *NC*-algorithms to compute determinants but there is a good chance (and our general belief) that there does not exist even a polynomial time *sequential* algorithm to compute permanents.

Recall that $\Lambda(n)$ denotes the number of processors needed to compute the determinant of an $n \times n$ matrix and the determinants of its minors in $O(\log^2 n)$ time. We stated in Chapter 5 that $\Lambda(n) = O(n^{3.5} \cdot m)$, if our numbers have $O(m \log(m))$ bits. We emphasize the importance of Λ since any improvement in the growth of Λ leads directly to an improvement in the parallel matching for general graphs.

In this chapter two basic algorithms are presented: one based on *partial-interpolation* and the second one using the *smallest-degree algorithm*.

The first one is simpler, but deals with integers of exponential size $O(2^m)$, so it needs $O(n^{3.5} \cdot m)$ processors, and it is $O(m)$ times less *efficient* (in the sense of processor bound) than the second one.

The second algorithm uses *modular arithmetics* in a clever way, thus reducing the size of the integers which are processed in the algorithm. Its efficiency (in the sense of the number of processors) is $O(n^{3.5})$. Hence it is essentially the same as the complexity of matrix inversion: the matching problem is reduced to the computation of the determinant of the whole matrix and its minors.

6.1 Constructing black-box polynomials

We start with the problem of testing the existence of a perfect matching in a given undirected graph G. The algebraic approach consists here in reducing this problem to a problem related to a *special black-box* polynomial; each monomial of such polynomials gives (at least one) perfect matching of the graph.

A *tour* of a square matrix A is a permutation $\sigma = (i_1, i_2, \ldots, i_n)$ such that all entries $A[1, i_1], A[2, i_2], \ldots A[n, i_n]$ are nonzero.

The value of the tour σ, denoted by product(σ), is the product

$$A[1, i_1] \cdot A[2, i_2] \cdot \ldots \cdot A[n, i_n].$$

We call it a *tour product.*

Denote by Tours(A) the set of all tours of a matrix A.

We introduce variables $x_{i,j}$ corresponding to the edges (i, j) of a given graph G. The assignment of variables to edges is done differently for general and for bipartite graphs. The situation is much simpler for the bipartite graphs and we start with this case.

Let $V1 = \{v_1 \ldots v_n\}$ and $V2 = \{w_1 \ldots w_n\}$. Assume $G = (V1, V2, E)$ is an equibipartite graph (which consists of some edges between the sets $V1$ and $V2$). Introduce variables $x_{i,j}$, $1 \leq i, j \leq n$.

The *Edmonds matrix* is the $n \times n$ matrix EDMONDS(G), whose (i, j)-th entry is $x_{i,j}$ if there is an edge (v_i, w_j), otherwise this entry equals zero.

Each tour corresponds to a potential matching $(1, i_1), (2, i_2), \ldots, (n, i_n)$. The following fact is obvious:

Lemma 6.1.1
The following conditions are equivalent:

The equibipartite graph G has a perfect matching;
Tours(EDMONDS(G)) $\neq \emptyset$;
det(EDMONDS(G)) *is a nonzero polynomial.*

A simple *RNC*-algorithm for bipartite perfect matching follows as a consequence of the fact that the zero test for the polynomial det(EDMONDS(G)) can be done in *RNC*, owing to the *nonzero-test theorem.*

Example Take the graph G illustrated in Figure 6.1. Its Edmonds matrix is

$$\text{EDMONDS}(G) = \begin{bmatrix} x_{1,1} & x_{1,2} & 0 \\ 0 & x_{2,2} & x_{2,3} \\ x_{3,1} & x_{3,2} & 0 \end{bmatrix}.$$

We have two tours of this matrix, $(2, 3, 1)$ and $(1, 3, 2)$, and the corresponding tour products are $x_{1,2}x_{2,3}x_{3,1}$ and $x_{1,1}x_{2,3}x_{3,2}$. The Edmonds polynomial is here

$$\det(\text{EDMONDS}(G)) = x_{1,2}x_{2,3}x_{3,1} - x_{1,1}x_{2,3}x_{3,2}.$$

The situation is more complicated for general undirected graphs (similarly as in the sequential computations). Assume now (for simplicity) that the set of vertices of G consists of numbers $1 \ldots n$. The correspondence between permutations $\sigma = (i_1, i_2, \ldots, i_n)$ and potential matchings of the graph is now more subtle.

In the case of bipartite matching each permutation corresponded to a potential perfect matching, but now only so-called *even-cycle permutations* correspond to potential perfect matchings.

For a permutation $\sigma = (i_1, i_2, \ldots, i_n)$ define its directed graph representation

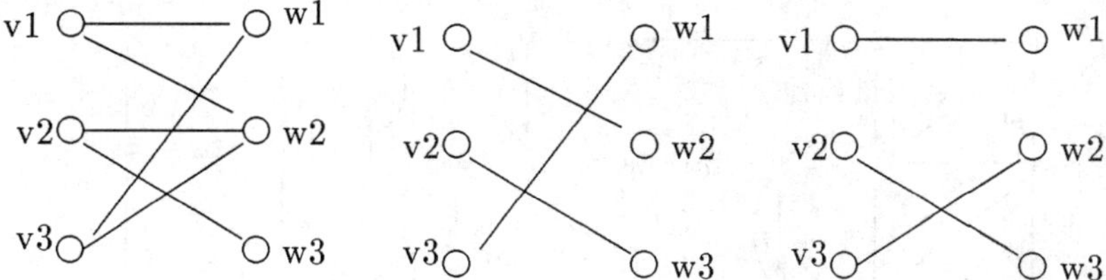

Fig. 6.1 An undirected bipartite graph G and all its perfect matchings.

$G' = \text{trail}(\sigma) = (V', E')$, where
$V' = [1 \ldots n]$ and $E' = \{(1, i_1), (2, i_2), \ldots (n, i_n)\}$.

Example The permutation $\sigma = (3412)$ corresponds to the graph consisting of two cycles:

$$G' = \text{trail}(\sigma)\colon 1 \leftrightarrow 3 \quad 2 \leftrightarrow 4.$$

The graphs corresponding in this way to permutations are called *trail graphs* here. Hence each trail graph is a directed graph whose node has its *indegree* and its *outdegree* equal to one. The graph consists of disjoint directed cycles.

We say that a permutation is *even-cycle* iff all cycles in its trail graph are of even length (in this case we say that its trail graph is even), otherwise the permutation is called *odd-cycle*.

Denote by ADJAC(G) the adjacency matrix of the graph G.

A trail subgraph of G is a trail graph with vertices $[1 \ldots n]$ whose edges (after removing their direction) are also the edges of G. Denote by TRAILS(G) the family of all trail subgraphs of G. An equivalent definition of the set of trail subgraphs is

$$\text{TRAILS}(G) = \{\text{trail}(\sigma)\colon \sigma \in \text{Tours}(\text{ADJAC}(G))\}.$$

If the graph has a perfect matching then by changing each edge of this perfect matching into a two-vertices cycle we get an even trail subgraph of G. On the other hand if we have an even trail subgraph of G then we can extract each second edge from each cycle. After removing the orientation these edges form a perfect matching of G.
This implies the following.

Fact The following conditions are equivalent:

an n-vertex graph G has a perfect matching;
there is an even trail graph $G' \in \text{TRAILS}(G)$;
there is an even-cycle permutation $\sigma \in \text{Tours}(\text{ADJAC}(G))$.

We say that a matrix A is a *skew-symmetric matrix* iff $A[i, j] = -A[j, i]$.

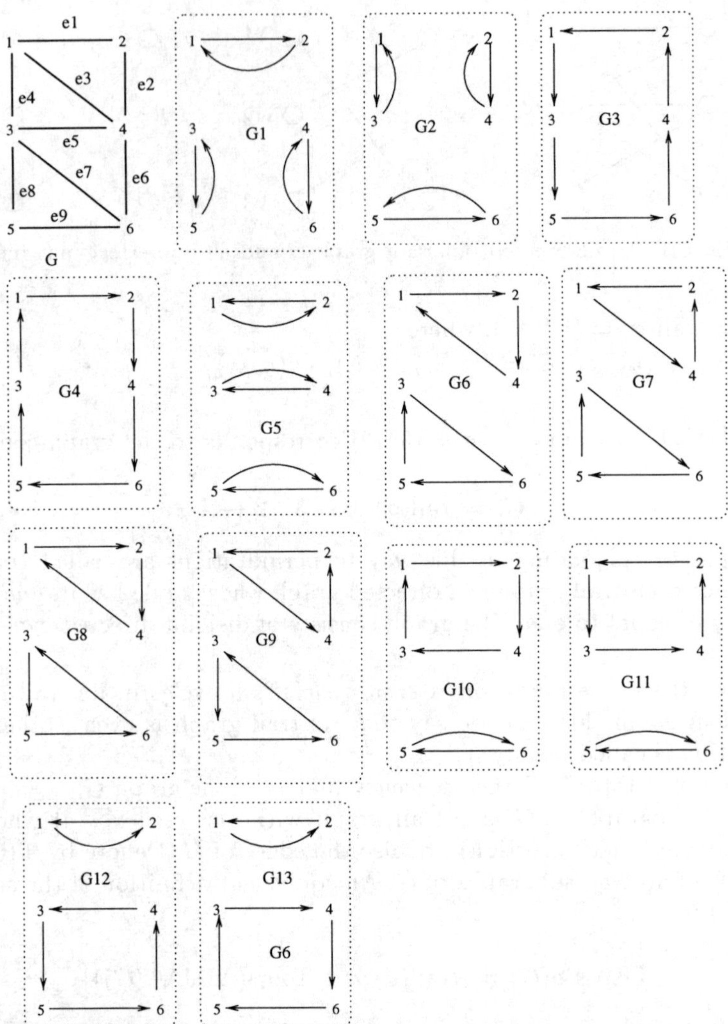

Fig. 6.2 An undirected graph G and the family TRAILS(G) of its directed trail subgraphs.

Introduce variables $x_{i,j}$, $1 \leq i < j \leq n$. Let AD = ADJAC(G). Define *Tutte's matrix* A of the graph G as a *skew-symmetric* $n \times n$ matrix A such that for each $1 \leq i < j \leq n$: $A[i,j] = x_{i,j}$ if $\text{AD}[i,j] \neq 0$. (Note that $A[i,i] = 0$ and $A[i,j] = -x_{ji}$ if $j <.i$ and $\text{AD}[i,j] \neq 0$.)

Definition
The basic algebraic object related to matchings in our book is the *Tutte symbolic polynomial* of a graph G:

$$\Psi(G) = \det(A). \tag{6.1}$$

It is our main *black-box* polynomial.

The weight of an edge (i, j) is the (i, j)-th entry of Tutte's matrix A. The weight of a permutation σ (written product(σ))is the product of the weights of the edges of the graph of the permutation σ.

Observation $\det(A) = \sum_\sigma \mathrm{sign}(\sigma) \cdot \mathrm{product}(\sigma)$ is a black-box polynomial.

We say that two permutations σ, τ are cyclic equivalent ($\sigma \equiv_c \tau$) iff τ results from σ by changing direction of a cycle of σ.

Changing direction of an edge corresponds to changing the sign of the weight of this edge. If a trail graph corresponding to a tour σ contains an odd cycle C then product(σ) is cancelled by product(τ), where τ results from σ by changing the direction of this odd cycle C.

This implies the first point of the lemma below, which is the key property of Tutte's matrix. Other points follow directly from definitions.

Lemma 6.1.2
Let A be Tutte's matrix of an undirected graph G.

1. *If σ is an odd-cycle tour of Tutte's matrix A, then*

$$\sum_{\sigma \equiv_c \tau} \mathrm{product}(\tau) = 0.$$

2. *If $\sigma \in$ Tours(A) and $\sigma \equiv_c \tau$ then also $\tau \in$ Tours(A).*
3. *If $\sigma \equiv_c \tau$ then* $\mathrm{sign}(\sigma) = \mathrm{sign}(\tau)$.

Example Let us consider the graph G illustrated in Figure 6.2. Tutte's matrix A of this graph is presented below.

$$A = \begin{bmatrix} 0 & x_{1,2} & x_{1,3} & x_{1,4} & 0 & 0 \\ -x_{1,2} & 0 & 0 & x_{2,4} & 0 & 0 \\ -x_{1,3} & 0 & 0 & x_{3,4} & x_{3,5} & x_{3,6} \\ -x_{1,4} & -x_{2,4} & -x_{3,4} & 0 & 0 & x_{4,6} \\ 0 & 0 & -x_{3,5} & 0 & 0 & x_{5,6} \\ 0 & 0 & -x_{3,6} & -x_{4,6} & -x_{5,6} & 0 \end{bmatrix}.$$

Example (cont.) In this example for $i < j$ we identify variables $x_{i,j}$ corresponding to edges of G with names of edges e_k according to Figure 6.2. So, for example, $x_{3,6} \equiv e_7$.

There are 13 tours $\sigma_1, \ldots, \sigma_{13}$ for the graph given in Figure 6.2. They correspond to the trail graphs shown on the figure. The tour products corresponding to them are

$$\begin{aligned}
&t_1 = -e_1^2 e_8^2 e_6^2;\\
&t_2 = -e_4^2 e_2^2 e_9^2;\\
&t_3 = -e_4 e_8 e_9 e_6 e_2 e_1;\\
&t_4 = t_3;\\
&t_5 = -e_1^2 e_5^2 e_9^2;\\
&t_6 = -e_8 e_7 e_9 e_3 e_2 e_1;\\
&t_7 = -t_6;\ t_8 = -t_6;\ t_9 = t_6;\\
&t_{10} = t_{11} = -e_1 e_4 e_5 e_2 e_9^2;\\
&t_{12} = t_{13} = -e_1^2 e_5 e_9 e_8 e_6.
\end{aligned}$$

The tours corresponding to graphs $G6, G7, G8, G9$ are odd, and cyclic equivalent. Their corresponding tour products sum to zero: $t_6 - t_6 - t_6 + t_6 = 0$.

Hence in our example Tutte's symbolic polynomial is

$$\begin{aligned}
\Psi(G) &= \det(A) = -t_1 - t_2 - t_3 - t_4 - t_5 + t_{10} + t_{11} + t_{12} + t_{13}\\
&= e_1^2 e_8^2 e_6^2 + e_4^2 e_2^2 e_9^2 + 2e_4 e_8 e_9 e_6 e_2 e_1 +\\
&\quad e_1^2 e_5^2 e_9^2 - 2e_1 e_4 e_5 e_2 e_9^2 - 2e_1^2 e_5 e_9 e_8 e_6.
\end{aligned}$$

The last lemma directly implies the following.

Lemma 6.1.3 ([T47], [LP86])
Let A be Tutte's matrix of G. Then G has a perfect matching iff Tutte's symbolic polynomial $\Psi(G) = \det(A)$ *is a nonzero polynomial.*

Observe that we could use polynomials without the signs of permutations, since

$$\mathrm{perm}(A) \neq 0 \Leftrightarrow \det(A) \neq 0.$$

The introduction of the signs of permutations could look artificial at first glance. However, this is the crucial trick, and it enables us to use determinants, which are easy to compute.

The main property of Tutte's symbolic polynomials can be stated as follows (it follows directly from the whole discussion above).

Theorem 6.1.4 (constructing black-box polynomials)
Let A be Tutte's matrix of a graph G and $\mathcal{M}$ *be the family of all perfect matchings of G. Then* $\Psi(G) = \det(A)$ *is an* $\mathcal{M}$*-consistent special black-box polynomial computable in* $\log^2 n$ *time with* $O(\Lambda(n))$ *processors in the sense of Chapter 5.*

The main result of this section now follows directly from Theorem 6.1.4 and Lemma 6.1.3 by the nonzero-test theorem from Chapter 5.

Theorem 6.1.5
The existence of a perfect matching for a given undirected graph G is in RNC.

Observation The construction of a perfect matching seems to be much harder then the existential test. Assume we have an NC-algorithm for testing existence of a perfect matching; it is an open question if this implies that construction of a perfect matching is in NC.

Remark Assume for each variable in the polynomial $\Psi(G)$ we substitute the square root of a different prime number. Then the value of the polynomial is nonzero iff the polynomial is nonzero. The only problem is to compute $\Psi(G)$ with the correct precision. Such an algorithm is a good candidate for an NC-algorithm testing existence of perfect matchings.

6.2 Bimatchings and Pfaffians

There is another ingenious algebraic concept related to matchings: the *Pfaffian* of a skew-symmetric square matrix A, denoted by PFAFFIAN(A). Let us fix a graph G in this section.Assume the set of nodes of G is $V = \{1, 2, \ldots, 2n\}$ and $\mathcal{M}$ is the set of all perfect matchings of G. If G is a complete graph then we can treat $\mathcal{M}$ as the set of all permutations of $[1 \ldots 2n]$ consisting of cycles of length 2 (so called 2-cycle permutations).

For $M \in \mathcal{M}$, $M \;=\; [(i_1, i_2), (i_3, i_4), \ldots (i_{2n-1}, i_{2n})]$ denote

$$\text{term}(M) \;= \text{sign}(i_1, i_2, \ldots, i_{2n}) \cdot A[i_1, i_2] \cdot A[i_3, i_4] \cdot \ldots \cdot A[i_{2n-1}, i_{2n}].$$

Observation If A is skew-symmetric then term(M) does not depend on the representation of M (changing order of blocks or within blocks).

Formally we define PFAFFIAN(A) for a $2n \times 2n$ Tutte matrix A of G, as follows:

$$\text{PFAFFIAN}(A) = \text{PFAFFIAN}(G) = \sum\nolimits_{M \in \mathcal{M}} \text{term}(M)$$

$$= \sum\nolimits_{[(i_1,i_2),(i_3,i_4),\ldots,(i_{2n-1},i_{2n})]} \text{sign}(i_1 i_2 \ldots i_n) \cdot A[i_1, i_2] \cdot A[i_3, i_4] \cdot \ldots \cdot A[i_{2n-1}, i_{2n}].$$

We show that there is a relation between Pfaffians and symbolic polynomials $\Psi(G)$ in terms of Equation 6.1:

$$\text{PFAFFIAN}(G)^2 = \Psi(G).$$

This is related to the fact that each term in $\Psi(G)$ is a sum of products of monomials corresponding to perfect matchings of G. We develop a useful representation of the symbolic polynomial Ψ in terms of collections of undirected cycles and edges.

A (perfect) *bimatching* $\mathcal{C}$ of G is an edge-union of two (not necessarily distinct) perfect matchings:

$$\mathcal{C} = M1 \cup M2, \text{ where } M1, M2 \in \mathcal{M}.$$

Denote by $\mathcal{BM}$ the set of all bimatchings of G. Observe that $\mathcal{BM} \neq \emptyset$ iff G has a perfect matching.

Observation Each bimatching $\mathcal{C}$ is a collection $\mathcal{C} = \{C_1, C_2, \ldots, C_k\}$ of disjoint undirected even-cycles and disjoint edges of G such that $\bigcup_{i=1}^{k} C_i = V$.

For a bimatching $\mathcal{C}$ denote by Directed_Versions($\mathcal{C}$) the set of all trails of G, which are directed versions of $\mathcal{C}$. Each such trail is generated from $\mathcal{C}$ in the following way: for each cycle of $\mathcal{C}$ all its edges are directed in the same direction to form a directed cycle, and each single edge component (u, v) is transformed into the cycle consisting of two edges $u \to v \to u$. Observe that two permutations are cyclic equivalent iff their trails are in the same set Directed_Versions($\mathcal{C}$).

For the graph in Figure 6.2 we have $\mathcal{M} = \{M1, M2, M3\}$, where

$$\begin{aligned} M1 &= \{(1,2),(3,5),(4,6)\}, \\ M2 &= \{(1,3),(2,4),(5,6)\}, \\ M3 &= \{(1,2),(3,4),(5,6)\}. \end{aligned}$$

For the matrix A corresponding to this graph we have

$$\text{PFAFFIAN}(A) = \text{sign}(123456) \cdot A[1,2] \cdot A[3,4] \cdot A[5,6]$$
$$+\,\text{sign}(123546) \cdot A[1,2] \cdot A[3,5] \cdot A[4,6] \;+\text{sign}(132456) \cdot A[1,3] \cdot A[2,4] \cdot A[5,6].$$

We have six bimatchings here:

$$M1 \cup M1, M2 \cup M2, M3 \cup M3,$$
$$M1 \cup M2, M1 \cup M3, M2 \cup M3.$$

In particular Directed_Versions($M1 \cup M3$) $= \{G12, G13\}$, see Figure 6.2.

Denote by $l(\mathcal{C})$ and $r(\mathcal{C})$ the number of all components and the number of components of $\mathcal{C}$ of size larger than 2 (components which are not isolated edges), respectively. Define product($\mathcal{C}$) = product(σ), where σ is the permutation corresponding to any trail in Directed_Versions($\mathcal{C}$). We know already that the product is the same for all cyclic-equivalent trails.

Observation

$$|\,\text{Directed\_Versions}(\mathcal{C})| = 2^{r(\mathcal{C})}.$$

If $\mathcal{D} \in$ Directed_Versions($\mathcal{C}$) then the sign of each permutation corresponding to $\mathcal{D}$ equals $(-1)^{l(\mathcal{C})}$.

Using this terminology we can represent the Tutte symbolic polynomial in the following form:

$$\Psi(G) \;=\; \sum_{\mathcal{C} \in \mathcal{BM}} (-1)^{l(\mathcal{C})} \cdot 2^{r(\mathcal{C})} \cdot \text{product}(\mathcal{C}) \qquad (6.2)$$

where the summation is over all bimatchings $\mathcal{C}$ of G.

We need the following simple lemma.

Lemma 6.2.1
Let q be even and let the permutation $\sigma' = (k_2, k_3, \ldots, k_q, k_1)$ be the cyclic shift of the permutation $\sigma = (k_1, k_2, \ldots, k_q)$. Then $\text{sign}(\sigma) \cdot \text{sign}(\sigma') = -1$.

Proof σ can be transformed easily to σ' with an odd number of transpositions. □

The following lemma is the key lemma for the relation of Pfaffians to polynomials $\Psi(G)$.

Lemma 6.2.2
1. Let $M1, M2 \in \mathcal{M}$ and a bimatching $\mathcal{C}$ be the union $M1 \cup M2$. Then

$$\text{term}(M1) \cdot \text{term}(M2) = (-1)^{l(\mathcal{C})} \cdot \text{product}(\mathcal{C})$$

2. For a given bimatching $\mathcal{C}$ we have

$$|\{(M1, M2) \colon M1, M2 \in \mathcal{M} \text{ and } \mathcal{C} = M1 \cup M2\}| = 2^{r(\mathcal{C})}.$$

Proof
1. Take any $D \in \text{Directed\_Versions}(\mathcal{C})$. We can write the cycles of D as

$$(i_1, i_2, i_3, \ldots, i_{t-1}, i_t),\ (j_1, j_2, \ldots, j_r),\ (l_1, l_2, \ldots, l_p) \ldots.$$

Then we can write $M1$, $M2$ as

$$[(i_1, i_2), \ldots, (i_{t-1}, i_t), (j_1, j_2), \ldots, (j_{r-1}, j_r), (l_1, l_2), \ldots, (l_{p-1}, l_p) \ldots].$$
$$[(i_2, i_3), (i_4, i_5), \ldots, (i_t, i_1), (j_2, j_3), (j_4, j_5), \ldots, (j_r, j_1),$$
$$(l_2, l_3), (l_4, l_5), \ldots, (l_p, l_1) \ldots].$$

It is clear that $\text{product}(\mathcal{C}) = \text{product}_1 \cdot \text{product}_2$, where
$\text{product}_1 = A[i_1, i_2] \cdot A[i_3, i_4], \ldots \cdot A[i_{t-1}, i_t] \cdot A[j_1, j_2] \cdot A[j_3, j_4], \ldots \cdot A[j_{r-1}, i_r] \cdot \ldots,$
$\text{product}_2 = A[i_2, i_3] \cdot A[i_4, i_5], \ldots \cdot A[i_t, 1] \cdot A[j_2, j_3] \cdot A[j_4, j_5], \ldots \cdot A[j_r, i_1] \cdot \ldots.$

Now it is enough to show that the product of signs of permutations corresponding to $M1$, $M2$ equals $(-1)^{l(\mathcal{C})}$. Howeve, these permutations are related through cyclic shifts of $l(\mathcal{C})$ subpermutations and the thesis follows from Lemma 6.2.1. This completes the proof of point (1).

2. We can compose a perfect matching M by choosing each second edge in each cycle to consist of more than one edge. M also contains all isolated edges of $\mathcal{C}$.

There are $r(\mathcal{C})$ cycles in $\mathcal{C}$ and therefore we have $2^{r(\mathcal{C})}$ possibilities to select one of two possible "*halves*" from each of them.

□

Lemma 6.2.2 directly implies the following fact.

Lemma 6.2.3
If $\mathcal{C} \in \mathcal{BM}$ *then*

$$\sum_{M1,M2 \in \mathcal{M}, M1 \cup M2 = \mathcal{C}} \mathrm{term}(M1) \cdot \mathrm{term}(M2) \;=\; (-1)^{l(\mathcal{C})} \cdot 2^{r(\mathcal{C})} \cdot \mathrm{product}(\mathcal{C})$$

Observe that the summation

$$\text{“}\textstyle\sum_{\mathcal{C} \in \mathcal{BM}} \sum_{M1,M2 \in \mathcal{M}, M1 \cup M2 = \mathcal{C}} \cdots \text{”}$$

is equivalent to the summation

$$\text{“}\textstyle\sum_{M1,M2 \in \mathcal{M}} \cdots \text{”},$$

and hence Equation (6.2) and Lemma 6.2.3 directly imply

$$\Psi(G) = \textstyle\sum_{\mathcal{C} \in \mathcal{BM}} (\sum_{M1,M2 \in \mathcal{M}, M1 \cup M2 = \mathcal{C}} \mathrm{term}(M1) \cdot \mathrm{term}(M2)) =$$

$$\textstyle\sum_{M1,M2 \in \mathcal{M}} \mathrm{term}(M1) \cdot \mathrm{term}(M2) = (\sum_{M \in \mathcal{M}} \mathrm{term}(M))^2 = \mathrm{PFAFFIAN}(A)^2.$$

The equation above implies the most important property of the Pfaffian which is its relation to the Tutte symbolic polynomial, see [LP86]. The crucial point is that $\det(A)$ can be treated as a symbolic polynomial $\Psi(\mathcal{M})$ of the family $\mathcal{M}$ of perfect matchings of G in the sense of equation (*) from Chapter 5.

Lemma 6.2.4
1. Let A be Tutte's matrix for a graph G. Then $(\mathrm{PFAFFIAN}(A))^2 = \det(A)$.
2. Let $\mathcal{M}$ be the family of perfect matchings of G, then

$$\det(A) = (\textstyle\sum_{M \in \mathcal{M}} a_M \cdot \mathrm{mono}(M))^2,$$

where $a_M = +1$ or $a_M = -1$ and $\mathrm{mono}(M)$ *is the product of variables corresponding to elements of M.*

A more formal algebraic proof of point 1 of Lemma 6.2.4 will be given in Chapter 13 (Corollary 13.2.3). The second point is a reformulation of the first one, since $\mathrm{mono}(M)$ is nonzero only for perfect matchings of G.

Example (cont.) For the graph G in Figure 6.2 only graphs $G1, G2, G5$ correspond to perfect matchings. Let A be Tutte's matrix of this graph. If we take only their corresponding terms together with signs of permutations then we obtain

$$\mathrm{PFAFFIAN}(A) = e_1 e_9 e_5 - e_2 e_4 e_9 - e_1 e_6 e_8.$$

Hence according to Lemma 6.2.4 we have

$$e_1^2 e_8^2 e_6^2 + e_4^2 e_2^2 e_9^2 + 2 e_4 e_8 e_9 e_6 e_2 e_1 + e_1^2 e_5^2 e_9^2 - 2 e_1 e_4 e_5 e_2 e_9^2 - 2 e_1^2 e_5 e_9 e_8 e_6$$

$$= (e_1e_9e_5 - e_2e_4e_9 - e_1e_6e_8)^2.$$

The relation between Pfaffians and Tutte's symbolic polynomial is especially visible for the following family of very simple graphs.

Example Assume G is the cycle C_n 1–2–3–...–n–1, where n is even. Let A be Tutte's matrix of C_n. Then Tutte's symbolic polynomial is

$$\det(A) = x_{12}^2x_{34}^2\dots x_{n-1,n}^2 \;+\; x_{23}^2x_{45}^2\dots x_{1,n}^2 \;+\; 2\cdot x_{12}x_{2,3}\dots x_{1,n}.$$

The Pfaffian of the same graph is

$$\text{PFAFFIAN }(A) = x_{12}x_{34}\dots x_{n-1,n} + x_{23}x_{45}\dots x_{1,n}$$

Then equality $\det(A) = \text{PFAFFIAN }(A)^2$ follows directly from the identity

$$(a+b)^2 = a^2 + b^2 + 2ab,$$

after substituting

$$a = x_{12}x_{34}\dots x_{n-1,n},\; b = x_{23}x_{45}\dots x_{1,n}.$$

If the coefficient of the monomial corresponding to each $M \in \mathcal{M}$ equals 1 then Pfaffian gives the exact number of perfect matchings, and it can be computed in *NC* by taking the square root of the determinant (we know that the determinant is a square of an integer number). This happens in the case of planar graphs.

Using Pfaffians one can count the number of perfect matchings in planar graphs in *NC*. However, existence of a corresponding *NC*-algorithm constructing perfect matchings in planar graphs is a difficult open problem.

6.3 Applying partial interpolation

We know that Tutte's symbolic polynomial is an $\mathcal{M}$-consistent special black-box polynomial, see Theorem 6.1.4. Hence it is enough to find any monomial of Tutte's symbolic polynomial that has a perfect matching of a given graph. We use the *partial-interpolation theorem* from Chapter 3 to do it. It is easy to see the following:

Theorem 6.3.1 (randomized-matching theorem)
We can find and test a perfect matching in a given graph (if it exists) by a randomized parallel algorithm working in $O(\log^2 n)$ time with $O(m\cdot n^{3.5})$ processors.

Proof We apply the partial-interpolation theorem. Let $\mathcal{P} = \det(A)$, where A is the Tutte's matrix of G. Then $\mathcal{P}$ is a special polynomial which is $\mathcal{M}$-consistent, where $\mathcal{M}$ is the family of all perfect matchings of G. The polynomials $\mathcal{P}_x(\bar{x}_0)$, for each variable x, can be easily computed using the minors of the matrix A. Alltogether we need to compute the determinants of all minors of A, and the determinant of A, for given values of variables. The whole computation can be done in $\log^2 n$ time with $O(n^{3.5}m)$ processors; an additional factor m is needed since the corresponding integers can have a linear number of bits.

□

Theorem 6.3.2 ([MVV87])
We can find a maximum cardinality matching in a given graph by a parallel randomized algorithm working in $O(\log^3 n)$ time with $O(m \cdot n^{3.5})$ processors.

Proof Assume that the size of the maximum matching in G is k. Then $n - 2k$ vertices are unmatched. Add $n - 2k$ new *special* vertices, and connect them to each vertex of G. Denote the constructed graph by $G(k)$. Then $G(k)$ has a perfect matching. On the other hand if $G(k)$ has a perfect matching then after removing all special vertices we receive a matching in G of size k. Hence, it is easy to see the following.

Claim Let

$$k_0 = \max\{k\colon\ G(k) \text{ has a perfect matching }\}.$$

Then the maximum cardinality of a matching of G is k_0.

Using the algorithm from Theorem 6.3.1 we can test if $G(2k)$ has a perfect matching for each k. We can use a binary search method (with respect to k) to find k_0. Then we find a perfect matching by the algorithm from Theorem 6.3.1. We remove special vertices and obtain a maximum cardinality matching in the original graph.

□

We can strengthen the theorem by producing an algorithm working in $O(\log^2 n)$ time. However, the algorithm becomes more complicated. A partial interpolation with weights is applied.

Theorem 6.3.3
(a) We can find a maximum cardinality matching in a given graph (if it exists) in $O(\log^2 n)$ time with $O(m \cdot n^{3.5})$ processors.
(b) Assume each edge of G has a weight given in unary. Then we can find the minimal weight perfect matching of G (if any exists) and maximum weight matching of G in $O(\log^2 n)$ time with $O(n^{3.5}m)$ processors.

Proof Apply the weighted partial-interpolation theorem to get point (b). Point (a) follows from (b) by adding new edges (to have the weighted complete graph G'). New edges get weight 1, and old ones get weight 0. The minimum weight perfect matching M in a new graph G' gives a maximum cardinality matching in G, by taking all edges from M of zero weight.

□

Remark Inclusion of the general maximum weighted perfect matching problem (for binary weights) in *RNC* is still an open problem, but the following partial result is known.

Theorem 6.3.4 ([Sp93])
For each constant k there is an RNC algorithm constructing a maximum weight matching in graphs with arbitrary weights with approximation factor $1 + \frac{1}{k}$.

6.4 Graphs with a small number of perfect matchings

Let $\mathcal{M}$ be the family of all perfect matchings of a given graph $G = (V, E)$. Let us fix G till the end of this section. Assume we know in advance that the number of perfect matchings of a graph G is bounded by a polynomial n^k, this means $|\mathcal{M}| \leq n^k$.

This knowledge enables us to test for existence and to compute all perfect matchings by an NC-algorithm.

Let $W \subseteq E$. Denote

$$\text{All\_Relevant}(W) = \{W \cap M \colon M \in \mathcal{M}\}.$$

We say that each subset $Y \in \text{All\_Relevant}(W)$ is *relevant* with respect to W.

Observation $|\text{All\_Relevant}(W)| \leq n^k$.

Lemma 6.4.1
Assume we know that the number of perfect matchings of a given graph G is bounded by n^k. Then for each Y, $W \subseteq E$ we can test the membership $Y \in$ All_Relevant(W) by a deterministic NC-algorithm.

Proof We can assume that $Y \subseteq W$ consists of *independent* edges (otherwise Y is not *relevant*). We construct an auxiliary graph. Remove from G all the edges of $W - Y$ without removing the endpoints of the removed edges from V. Then remove all edges of Y together with their endpoints. Denote the resulting graph by G'. Then the membership $Y \in \text{All\_Relevant}(W)$ is equivalent to the fact that G' has a perfect matching. Moreover, each perfect matching M' of G' can be extended to the perfect matching $M' \cup Y$ of G. Hence G' has polynomially many perfect matchings. We can test for a perfect matching in G' by an NC-algorithm due to the *sparse-test theorem* from Chapter 5.

□

Theorem 6.4.2 (sparse-matching theorem, [GK87])
Let $\mathcal{M}$ be the family of all perfect matchings of G. If $|\mathcal{M}|$ is bounded by a known polynomial then we can compute $\mathcal{M}$ by an NC-algorithm.

Proof Assume, without loss of generality, that the number of edges m is a power of two. Partition the set E into two sets $E1$, $E2$ of the same cardinality. Then partition these two sets into halves etc. We receive the family $\mathcal{S}$ consisting of a linear number of subsets of E. Assume W is one of these subsets. Denote by FirstHalf(W) and SecondHalf(W) the disjoint subsets of W which are in $\mathcal{S}$ (of the same cardinality).

Let $\Delta(W)$ denote the family of all maximal (with respect to inclusion) subsets of W which are relevant.

For two families $\mathcal{F}_1$, $\mathcal{F}_2$ of subsets of E denote

$$\mathcal{F}_1 \otimes \mathcal{F}_2 = \{Y_1 \cup Y_2 : Y_1 \in \mathcal{F}_1, Y_2 \in \mathcal{F}_2\}.$$

The function All_Relevant(W) computes $\Delta(W)$ recursively for each subset of edges $W \in \mathcal{S}$. Due to the small number of perfect matchings all computed families All_Relevant(W) are of polynomial size. Observe that All_Relevant(E) $= \mathcal{M}$. □

```
function All_Relevant(W); {assume |W| is a power of 2}
if |W| = 1 {W = {e} for some edge e = (v, w)} then
    begin
    let G' be G with nodes v, w removed;
    if G' has a perfect matching then return {W}
        else return ∅;
    end
else
    begin
    {partition W into two equal subsets}
    W1 := FirstHalf(W); W2 := SecondHalf(W);
    for i ∈ {1,2} do in parallel
            F_i := All_Relevant(W_i);
    F := F1 ⊗ F2;
    for each Y ∈ F do in parallel
            if Y ∉ All_Relevant(W) then remove Y from F;
    return F;
    end
```

Remark Applying the sparse-test theorem from Chapter 5 for Tutte's symbolic polynomial as the black-box polynomial we can test if a graph has a perfect matching. In the algorithm powers of prime numbers are used. This algorithm is a good candidate for an *NC*-algorithm for testing perect matchings in the general case (possibly after slight modifications). It is quite probable that we can perform the algorithm using only constant number of powers of prime numbers (if polylogarithmically many powers suffice then we also get an *NC*-algorithm).

6.5 From Monte Carlo to Las Vegas

The interpolation-based algorithms given before are *Monte Carlo* algorithms. They can sometimes give a wrong answer, though the probability of that is astronomically small. The algorithm outputs a maximum matching of size k, where $k \leq \text{MatchNum}(G)$ and with high probability $k = \text{MatchNum}(G)$. Anyway the answer could be incorrect.

We now give a *Las Vegas* algorithm for the maximum matching problem; the algorithm is called *Las Vegas* iff its *final* output is always correct. In our case the expected time is of the same order as before (i.e. $O(\log^2 n)$).

We apply the combinatorics of matchings developed in Chapter 2. Let $G = (V, E)$. Recall that for $X \subseteq V$ we defined $G - X$ as the graph which results after deleting all nodes of X from G, together with their incident edges, and $\theta(G')$ is the number of connected components of G' having odd cardinality.

A set X which maximizes $|\theta(G-X)| - |X|$ is called (in Chapter 2) a *Tutte set* of G. Recall that a vertex $v \in V$ is *critical* iff $\mathrm{MatchNum}(G) > \mathrm{MatchNum}(G - v)$. In other words v participates in each matching of G of maximum cardinality.

In Chapter 2 we defined

$$\mathrm{TutteSet}(G) = \{\, v \in V\colon\ v \text{ is critical and has a noncritical neighbor}\}.$$

We also presented in Chapter 2 the *Berge formula* which says essentially that we can compute the matching number knowing the set which maximizes Local_Defect. The Tutte set satisfies this condition, see the witness theorem in Chapter 2. We can compute the matching number as follows:

$$\mathrm{MatchNum}(G) = \frac{n - (|\theta(G - \mathrm{TutteSet}(G))| - |\,\mathrm{TutteSet}(G)|}{2}.$$

By the Berge formula, it is enough to find a Tutte set to compute the matching number. All components, including odd ones can be computed in *NC* efficiently.

The Tutte set $S = \mathrm{TutteSet}(G)$ could be found if we compute all critical vertices (owing to the definition of $\mathrm{TutteSet}(G)$). It is enough to compute all numbers $\mathrm{MatchNum}(G - v)$ for $v \in V$. We can use the algorithm from the *randomized-matching* theorem. This algorithm is a Monte-Carlo-type algorithm.

Once we have computed a Tutte set S then we can compute $\mathrm{MatchNum}(G)$. Denote by $\mathrm{MatchNum}\,1(G)$ the output of such an algorithm. Then

$$\mathrm{MatchNum}\,1(G) \geq \mathrm{MatchNum}(G) \text{ and}$$
$$Prob\{\mathrm{MatchNum}\,1(G) = \mathrm{MatchNum}(G)\} \text{ is high.}$$

On the other hand the result of the algorithm from the randomized-matching theorem, denoted here by $\mathrm{MatchNum}\,2(G)$, satisfies the following conditions:

(1) $\quad \mathrm{MatchNum}\,2(G) \leq \mathrm{MatchNum}(G)$ and

(2) $\quad Prob\{\mathrm{MatchNum}\,2(G) = \mathrm{MatchNum}(G)\}$ is high.

Observe that:

if $\mathrm{MatchNum}\,1(G) = \mathrm{MatchNum}\,2(G)$ then $\mathrm{MatchNum}\,1(G) = \mathrm{MatchNum}(G)$.

Now we can use both algorithms, and output the result only if

$$\text{MatchNum}\,1(G) = \text{MatchNum}\,2(G).$$

This happens with high probability, and we are sure that the answer (if there is one) is correct.

The algorithm from the randomized-matching theorem not only computes MatchNum 2(G) but also outputs a matching of this size. This proves the following theorem.

Theorem 6.5.1 (Las Vegas matching theorem, [K86])
There is an RNC Las Vegas algorithm to find a maximum cardinality matching.

6.6 The shrinking algorithm

In this section we present a different approach to construct a perfect matching of a graph (if there is any). It consists in successive shrinking by a given graph until there are no more edges. The algorithm is more complicated but its advantage is the small size of the numbers (observe that the main trick in the previous algorithm was an implementation of the isolating theorem with the use of very large numbers).

Assume that a graph G has a perfect matching and its number of edges satisfies $m \geq 3/4n$. We show how to remove a large proportion of the edges of G in a *safe* way, which means that after application of this procedure the graph still has a perfect matching. This operation shrinks a graph. Eventually, after several such shrinkings, we obtain a (sparse) graph with at most $3/4n$ edges.

Then a (rather simple) different shrinking procedure is applied to a sparse graph to compute a large partial matching M'. This matching is included in the final matching and removed from the graph. We are certain, with large probability, that a smaller graph (after each of two types of shrinking) has a perfect matching. We iterate until the graph is "shrunk" to an empty graph (no edges).

The main point is how to find a large subset of edges whose removal is *safe* with respect to the existence of a perfect matching. We introduce the concepts of the *rank* and *redundant* edges. The redundant edges are safe with respect to removal.

Let $\mathcal{M}$ be the family of all perfect matchings of a given graph G. For $S \subseteq E$ define

$$\begin{aligned}\text{rank}(S) &= \max\{|S \cap M| \;:\; M \in \mathcal{M}\}\\ \text{redundant}(S) &= \{e \in E - S \;:\; \text{rank}(S) = \text{rank}(S \cup \{e\})\}\end{aligned}$$

The concept of redundant edges is quite subtle. It is possible that an edge e is redundant with respect to S but not redundant with respect to some set $S \cup \{e'\}$. This happens for example in Figure 6.3, if we take $S = \{(1,2)\}$, $e = (1,5)$ and $e' = (2,3)$.

It follows directly from the definition of redundant edges that their removal is safe. If $\text{rank}(S) = |S \cap M|$ for a perfect matching M then $\text{redundant}(S) \cap M = \emptyset$, so after removing all edges in redundant(S) the set M is still a perfect matching in a shrinked graph.

Lemma 6.6.1 (safeness lemma)
Let $S \subseteq E$, then if G has a perfect matching then after removing all edges in redundant(S) the graph G has still a perfect matching.

```
Algorithm Perfect_Matching_By_Shrinking;
while G ≠ ∅ do
begin if G has an isolated vertex then failure;
    Let n', m' be the actual number of vertices (edges);
    if m' ≥ (3/4)n' then
    begin
        S := a random subset of V of size 1 ≤ i ≤ (2/3)m';
        R := redundant(S); {the most costly operation}
        G := G − R;
    end else
    begin
        let M' be the set of edges incident to a vertex of degree 1;
        M := M ∪ M'; G := G ⊖ M';
    end;
end;
 if M is a perfect matching of the initial graph then return M;
```

Assume that if R is a set of edges then the operation $G \ominus R$ removes from G all edges in R together with their endpoints. This implies that some other edges are also removed in this operation because one of their endpoints disappears. On the other hand the operation $G - R$ here removes only the edges from R, without removing their endpoints.

Example Consider the graph in Figure 6.3. Let $S1 = \{(1,2),(2,3)\}$. In the initial graph we have

$$\text{rank}(S1) = 1,\ \text{redundant}(S1) = \{((2,6),(2,5),(2,4),(1,5),(3,5)\}.$$

The edge $(4,5)$ is not redundant. For a perfect matching $M = \{(1,2),(3,6),(4,5)\}$ we have

$$|(S1 \cup \{(4,5)\}) \cap M| = 2 > \text{rank}(S1) = 1.$$

If we apply the algorithm Perfect_Matching_By_Shrinking to the initial graph in Figure 6.3(A) then we obtain the graph (B). Then we take $S = \{(6,3)\}$, remove redundant edges w.r.t. S and obtain the graph shown in (C). The algorithm stops and returns the matching

$$M = \{(1,2),(3,6),(4,5)\}.$$

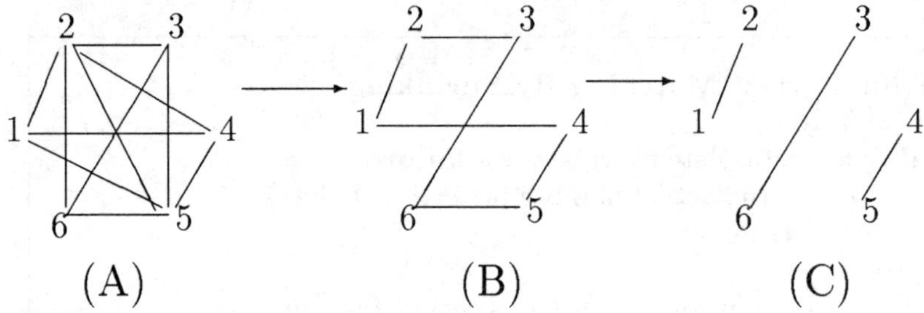

Fig. 6.3 A possible history of the shrinking algorithm. In the first shrinking we take $S = \{(1,2),(2,3)\}$, and in the second shrinking we take $S = \{(6,3)\}$.

Lemma 6.6.2 (analysis of the shrinking algorithm)
The average number of iterations of the algorithm Perfect_Matching_By_Shrinking *is* $O(\log n)$.

Proof From the *large-redundancy theorem* from Chapter 2 we have the following implication:

> if we pick a random subset S of edges of cardinality at most $\frac{2}{3}m$ then $Prob\{|\,\text{redundant}(S)| \geq m/72\} \geq 1/72$.

Hence after each application of the operation redundant the graph shrinks by a factor $(1 - 1/72)$ with high probability.

If the number of edges is smaller than $\frac{3}{4}n$ than there are at least $n/2$ vertices of degree 1. Then also in this case, in each iteration a large proportion of edges is deleted from the graph with high probability. Hence the average number of iterations is logarithmic.

□

6.7 Application of the smallest-degree algorithm

We come now to the main task in the algorithm Perfect_Matching_By_Shrinking: computation of the set redundant(S). Assume that a given graph G has a perfect matching. The edge e is redundant with respect to S iff $\text{rank}(S) = \text{rank}(S \cup \{e\})$. Hence we only have to compute the ranks of all sets of the form $S \cup \{e\}$ for $e \in E - S$.

We show how to compute rank(S) and relate it to the smallest degree of certain polynomials. We start with bipartite graphs, which are much easier. Let A = EDMONDS(G) be the Edmonds matrix of an equibipartite graph of order n. Assume the graph has a perfect matching. Let us modify the matrix by introducing a new variable z.

For each edge $(i, j) \notin S$ change the entry $A[i, j] = x_{i,j}$ to $A[i, j] = z \cdot x_{i,j}$. Denote the resulting matrix by A_S.

Lemma 6.7.1 (bipartite-rank lemma)
Assume G has a perfect matching. Then $\text{rank}(S) = n - \text{mindeg}_z(\det(A_S))$.

Proof Each even-cycle tour of the matrix A corresponds to a perfect matching and a monomial of Tutte's symbolic polynomial $\det(A)$. Assume $r = \text{rank}(S) = |M \cap S|$, where M is a perfect matching. The monomial corresponding to M has degree $n - r$ with respect to z, since when its corresponding tour "traverses" an edge (i, j) then the contribution to the degree (with respect to z) is one if $(i, j) \notin S$, otherwise it contributes zero to the degree of z.

The total length of each tour of the matrix is n. Hence the smallest degree of the resulting monomial with respect to z equals n minus the number of times the tour "traverses" through edges of S, so it is $n - |M \cap S|$.

□

Recall that

$$\text{redundant}(S) = \{e \in E - S \ : \ \text{rank}(S) = \text{rank}(S \cup \{e\})\}.$$

Hence, if we are to compute redundant(S) according to this definition, we have to compute rank(S) and rank($S \cup \{e\}$) for each edge $e \notin S$ of G. By the bipartite-rank lemma this can be done by computing degrees $\text{mindeg}_z(\det(A_S))$ and $\text{mindeg}_z(\det(A_{S\cup\{e\}}))$ for each $e \in E - S$.

The smallest degrees are computed by substituting random values for the variables, according to the algorithm for smallest degrees from Chapter 5.

A quadratic number of edges e is possible, so it seems that we need to calculate a quadratic number of determinants $\det(A_{S\cup\{e\}})$. However, the best known methods to compute the determinant $\det(A_S)$ also compute determinants of all its minors. Recall that the minor $A_{i,j}$ is a matrix which results from A by removing its i-th row and its j-th column.

Observe that the determinant $\det(A_{S\cup\{e\}})$ differs from $\det(A_S)$ in only one entry. We use the following algebraic lemma, which shows that once we have computed determinants of minors of A_S, we do not need to compute each of the determinants $\det(A_{S\cup\{e\}})$ from scratch.

Lemma 6.7.2 (single-change lemma)
Assume we know $\det(A)$ *and the determinants of all minors $A_{i,j}$ of an $n \times n$ matrix A. Let A' be a matrix which results by changing one entry of A. Then* $\det(A')$ *can be computed sequentially (by one processor) in a constant time.*

Proof Assume we have changed a single entry $a_{i,j}$ to a new value $a'_{i,j}$. Then, from the Laplace formula for determinants, we have

$$\det(A') = \det(A) + (a_{i,j} - a'_{i,j}) \cdot \det(A_{i,j}).$$

Hence $\det(A')$ can be computed in constant time.

□

Assume the determinant of the matrix and the determinants of all minors of the matrix can be computed in $\log^2 n$ time with $P(n)$ processors. Hence for bipartite graphs, by Lemma 6.7.2, we can compute the set redundant(S) within the complexities of the same order.

The situation is much more complex for a general graph. We want to use a similar algebraic lemma (single-change lemma), but using a direct approach we obtain matrices which differ in two entries. Nevertheless, we shall be able to use the trick from Lemma 6.7.2.

We explain now how to compute redundant(S) for general graphs.

Let B be the Tutte's matrix of a graph G. Let us modify the Tutte's matrix B by introducing a new variable z. Then for each edge $(i,j) \notin S$ multiply $B[i,j]$ by z. Denote the resulting matrix by B_S.

Let M be a perfect matching realizing maximal value of $r = |M \cap S|$. Take a tour corresponding to M whose trail consists of cycles of size 2. Hence each edge of M is traversed in this tour twice. Essentially the same argument as in the proof of the bipartite-rank lemma can be used to prove point (1) of the lemma below. Point (2) follows directly from point (1).

Lemma 6.7.3 (general-rank lemma)
(1) *Assume* G *has a perfect matching. Then* $2\cdot\text{rank}(S) = 2\cdot n - \text{mindeg}_z(\det(B_S))$.
(2) $e \in \text{redundant}(S)$ *iff* $\text{mindeg}_z(\det(B_S)) > \text{mindeg}_z(\det(B_{S\cup\{e\}}))$.

The matrix $B_{S\cup\{e\}}$ differs from B_S in exactly two entries (i,j) and (j,i), where $e = (i,j)$, for some $1 \le i < j \le n$. The variable z does not appear in these entries. This means that

$$B_S(i,j) = z \cdot B_{S\cup\{e\}}(i,j),$$

$$B_S(j,i) = z \cdot B_{S\cup\{e\}}(j,i).$$

Unfortunately $B_{S\cup\{e\}}$ differs from B_S in two entries, and we cannot apply the single-change lemma directly. We overcome this difficulty by introducing matrices D_e which are "better" than $B_{S\cup\{e\}}$, in the sense of applicability of Lemma 6.7.2.

Let us fix a subset $S \subseteq [1 \ldots n]$. To reduce the notation we write B instead of B_S. We define

$$D_e = B_e^+ + B_e^-,$$

where B_e^+ and B_e^- are defined as follows.

B_e^+ is the same as B, except for the (i,j)-th entry which is set as $B_e^+(i,j) = x_{i,j}$ (while $B(i,j) = z \cdot x_{i,j}$). Similarly B_e^- is the same as B, except for the (j,i)-th entry, where $B_e^-(j,i) = x_{j,i}$.

We omit the algebraic proof of the following lemma, which is a version of Lemma 6.7.3:

Lemma 6.7.4
$e \in \text{redundant}(S)$ *iff* $\text{mindeg}_z(\det(B)) > \text{mindeg}_z(\det(D_e))$.

The main result of this section is summarized by the following theorem.

Theorem 6.7.5 ([GP88])
A perfect matching in a general undirected graph (if there is any) can be found in $\log^3 n$ *randomized time with* $O(n^{3.5})$ *processors.*

Proof Observe that

$$\text{mindeg}_z(\det(D_e)) = \min\{\text{mindeg}_z(\det(B_e^+)), \text{mindeg}_z(\det(B_e^-))\}.$$

The crucial point, to reduce complexity, is that each of the matrices B_e^+ and B_e^- differs from B only in a single entry, though D_e differs in two entries.

The smallest degrees of $\det(B_e^+)$ and $\det(B_e^-)$ are computed by substituting random values for the variables, according to the algorithm for smallest degrees from Chapter 5. For a given assignment of variables, the matrices become integer matrices. We can calculate the determinant of B and the determinants of all minors of B. Then all determinants $\det(B_e^+)$ and $\det(B_e^-)$ for all $e \notin S$ can be easily calculated from the determinants of minors, using Lemma 6.7.2.

We need a logarithmic number of iterations. Hence $O(\log^3 n)$ time and $O(\Lambda(n))$ processors are enough.

We know that $\Lambda(n) = O(n^{3.5})$.

□

6.8 Bibliographic notes

The algorithm using partial interpolation is essentially from Mulmuley , Vazirani and Vazirani [MVV87] and the one using smallest degrees is from Galil and Pan [GP88]. The transformation to the Las Vegas algorithm is from Karloff [K86]. The matching algorithm for graphs with a small number of perfect matchings is from Grigoriev and Karpinski [GK87].

7

Inclusion maximal matchings

In this chapter we consider matchings M which are maximal with respect to inclusion. This means that in a given graph G there is no matching $M' \neq M$ such that $M \subset M'$. If X is a set of vertices then denote by Incident(X) the set of edges incident to any vertex in X. Denote also the set of edges incident with any endpoint of an edge in M by Incident(M); hence $M \subseteq$ Incident(M).

Observation Let M be a set of independent edges, and X be the set of endpoints of edges in M. Then the following conditions are equivalent:

(a) M is an inclusion maximal matching.
(b) $E =$ Incident(M).
(c) $E =$ Incident(X).

Example Let $G = P_3$, where $P_4 = (v_1, v_2, v_3, v_4)$ is the simple path v_1–v_2–v_3–v_4 of length 4. If we take $M = \{(v_2, v_3)\}$ then M is an inclusion maximal matching. Obviously M is not a maximum cardinality matching, since the matching $M' = \{(v_1, v_2), (v_3, v_4)\}$ has larger cardinality equal to 2.

We see in this example that the cardinality of M can be two times smaller than the cardinality of the maximum matching. It happens that this is the *worst case*.

Lemma 7.0.1
Assume M' is a maximum cardinality matching and M is an inclusion maximal matching. Then $|M'|/|M| \leq 2$ and $|E|/|M| \leq 2$.

Proof The first inequality follows trivially from the second one. Let V' be the set of nodes *matched* by M. It is easy to see that for each edge $(v, w) \in E$ we have $v \in V'$ or $w \in V'$. Hence $|V'| \geq |E|$. Now the inequality follows from the fact that $|M| = |V'|/2$.

□

7.1 Deterministic algorithm for maximal matchings

The first approach to construct an NC-algorithm for the inclusion maximal matching problem would be to implement a sequential algorithm in parallel .

There is an extremely simple linear time sequential algorithm which we call the *greedy algorithm*. The algorithm SEQ_GREEDY returns a maximal matching M for a given graph $G = (V, E)$.
Unfortunately the *exact* simulation of the greedy algorithm is not known to be well parallelizable.

```
Algorithm SEQ_GREEDY;
{returns a maximal matching M}
M := ∅;
while E ≠ ∅ do
    begin
    choose any edge e ∈ E
    M := M ∪ {e};
    E:= E – Incident(e);
    end;
return M;
```

Although the algorithm SEQ_GREEDY is not directly parallelizable, its main idea works well in parallel. However, in the parallel case, in a single parallel step, we do not delete single edges but larger groups PartialMatch$(G) \subseteq E$ of many independent edges in G; we remove (in parallel) all edges incident to the selected ones.

```
Algorithm PARALLEL_GREEDY;
M := ∅;
while E ≠ ∅ do
    begin
    X := PartialMatch(G)
    M := M ∪ X;
    E:= E – Incident(X);
    end;
return M; {M is a maximal matching}
```

There are two similar NC-algorithms for the computation of inclusion maximal matchings: the first (the earlier one) due to Israeli and Shiloach [IS84], the second one due to Kelsen [K94]. Both algorithms use the technique of Euler partitioning. We give a version of Kelsen's algorithm.

Observation Assume that at each stage in the algorithm PARALLEL_GREEDY we select the set X = PartialMatch(G) such that $|\text{Incident}(X)| \geq |E|/6$. Then the number of iterations of the algorithm is logarithmic.

Our main goal is to construct an NC-algorithm which computes PartialMatch(G) and satisfies the inequality in the observation above.

The main tool in the design of PartialMatch is a technique called *Euler partitioning*, see [GR88]. The technique consists in decomposing the set of edges of the graph into disjoint cycles and paths. The resulting decomposition, written Euler_Partition(G), has the following property for each vertex v:

(*) v is an endpoint of at most one open path in Euler_Partition(G).

The technique of Euler partitioning is much easier for bipartite graphs, so assume for a little time that G is bipartite (we drop this assumption soon). For each vertex v we can make a kind of local *pairing* of edges incident to v. If $\deg_G(v)$ is odd then denote by Incident$'(v)$ all incident edges except one. If $\deg_G(v)$ is even then Incident$'(v)$ = Incident(v).

Decompose Incident$'(e)$ into the set of disjoint pairs: $(e_i, e_i') : 1 \leq i \leq \lfloor \deg_G(v)/2 \rfloor$.

Define Next$(e_i) = e_i'$ and Next$(e_i') = e_i$. The function Next is constructed in parallel for all vertices v. It can happen that for some edge e the function NEXT(e) is undefined in this moment (it was never used in some pairing of edges). Then e is treated as a one-edge path.

The function Next, together with some single edges, gives a decomposition Euler_Partition(G) of the set of all edges into disjoint paths and cycles satisfying (*). The decomposition Euler_Partition(G) has property (*).

Remark For bipartite graphs the partitioning Euler_Partition(G) can even be done by an optimal algorithm (the product *time* $\times$ *processors* is linear with respect to the size of the graph). The improvement is rather technical, see [K94].

The main operation in our algorithm is the operation of *halving* degrees of the vertices.

```
function Halve(G);
D := Euler_Partition(G);
choose a unique orientation for each cycle and each path in D;
remove each second edge from each cycle and path of D;
{if a path is a single edge then do not remove it}
return the set of remaining edges;
```

Lemma 7.1.1 (halving degrees)
Let G' = Halve(G). *Then* $\lfloor \deg_G(v)/2 \rfloor \leq \deg_{G'}(v) \leq \lceil \deg_G(v)/2 \rceil$. Halve($G$) *can be computed in* $O(\log n)$ *time with* $O(n)$ *processors.*

The operation Halve is the basic component in the construction of a forest (called KelsenForest) whose nonleaf vertices have sufficiently large total sum of their degrees.

```
function KelsenForest(G);
    i := 0;
while E ≠ ∅ do
    begin
(#)     H := {v : deg_G(v) = 1} ;
        {let level(v) = i for each v ∈ H }
        for each v ∈ H do in parallel
                father(v) := the unique neighbor w of v;
                {if deg(w) = 1 then exactly one endpoint of the edge
                (v, w) is chosen as the father of the second one}
        E := E − Incident(H); V := V − H;
        G: = Halve(G); i : = i + 1
    end;
```

Remark The concept of KelsenForest does not appear explicitly in [K94].

Let $\mathcal{F}$ be a *forest* (collection of *rooted trees*). Denote by MaxForestMatch($\mathcal{F}$) the operation which computes a matching in $\mathcal{F}$ such that each nonleaf vertex of the forest is *matched*. We omit here the proof of the following lemma, it can be done by applying any *tree-contraction* technique.

The forest consists of trees, which form an easy subclass of graphs with respect to matchings (see Chapter 9), and the maximum matching algorithm from Chapter 9 can be applied to each tree in the forest. In fact we need the lemma later for the case when the trees of the forest are of logarithmic height; then the proof is much simpler (going top down, level by level).

Lemma 7.1.2 (forest-matching)
MaxForestMatch($\mathcal{F}$) *can be computed in* $O(\log n)$ *time with* $O(n)$ *processors.*

Example Consider the graph G presented in Figure 7.1. Assume that the Euler partition is as shown in the figure. The first application of the operation Halve removes each second edge (dotted edges in the figure). Then we obtain the graph Halve(G).

Observe that the constructed partial matching $M' = \text{MaxForestMatch}(\mathcal{F})$ is not inclusion maximal. However, after removing M' and all incident edges (in the Kelsen Algorithm) only two edges $(5,7)$, $(6,8)$ survive. These edges are added to the final matching:

$$M = \{(1,2),(3,4),(5,7),(6,8),(9,10),(11,12)\}.$$

M is the result of the Kelsen Algorithm, and is inclusion maximal. Accidently, it is also of maximum cardinality.

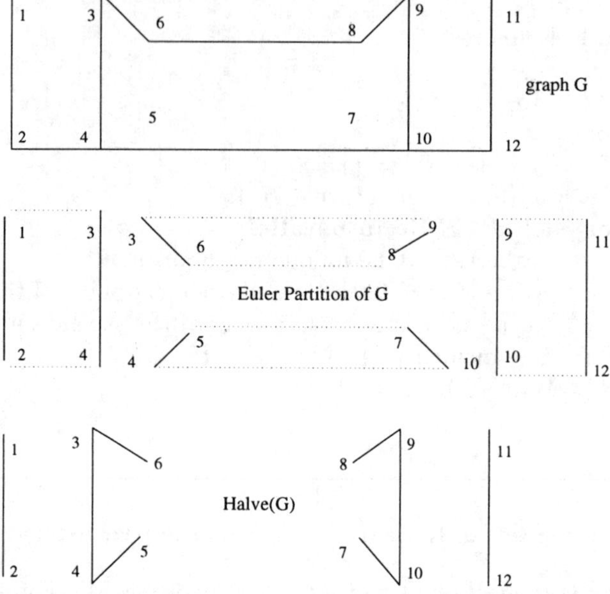

Fig. 7.1 An undirected graph G, its Euler partition and the graph Halve(G). Each dotted edge is removed from G.

Assume that the operation KelsenForest(G) does not change the set of edges; the operation can be modified by restoring the original edges, or assuming that it works on set of edges *local* to the function.

The version of the Kelsen algorithm presented here iteratively finds forests $\mathcal{F} = \text{KelsenForest}(G)$ and partial matchings MaxForestMatch$(\mathcal{F})$. The number of iterations is logarithmic due to Lemma 7.1.3.

Kelsen Algorithm;
$M := \emptyset$;
while $E \neq \emptyset$ **do**
  begin
    $\mathcal{F} :=$ KelsenForest(G);
    $M' :=$ MaxForestMatch$(\mathcal{F})$;
    $M := M \cup M'$;
    $E := E -$ Incident(M');
  end;
return M;

Lemma 7.1.3 (key-lemma)
Let $\mathcal{F} = \text{KelsenForest}(G)$, $M' = \text{MaxForestMatch}(\mathcal{F})$. *Then* $|\text{Incident}(M')| \geq |E|/3$.

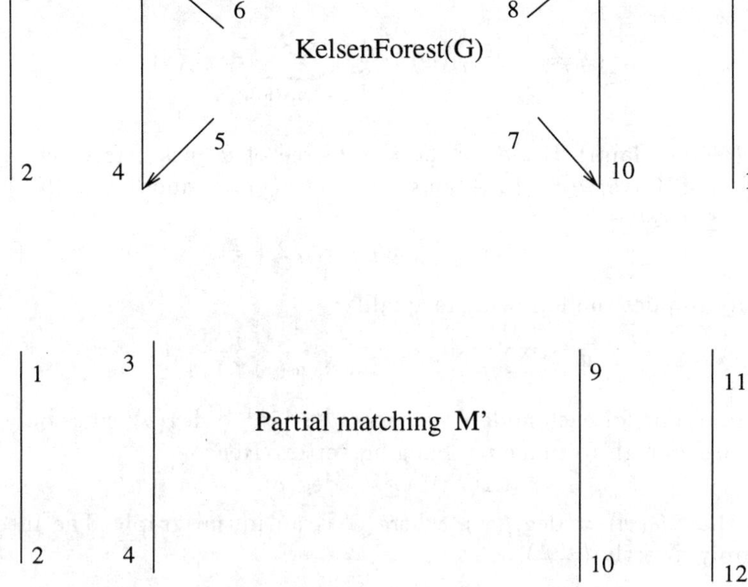

Fig. 7.2 $\mathcal{F}$ = KelsenForest(G) and M' = MaxForestMatch($\mathcal{F}$), where G is the graph from Figure 7.1. We assume that in the case of *ties* (in the moment when two vertices have degree 1 and they are neighbors) the larger vertex is a son of a smaller one.

Proof Observe that $\mathcal{F}$ has no isolated vertices and $\mathcal{F}$ *spans* the whole graph. Denote by Non_Leaves($\mathcal{F}$) the set of nonleaf vertices of $\mathcal{F}$. The following claim follows easily from the definition of MaxForestMatch.

Claim A
Each vertex in Non_Leaves($\mathcal{F}$) is *matched* by M'.

Assume that the initial value of the variable i is set to zero before the initial call to KelsenForest(G), i is the counter of performed operations Halve and it is not really needed in the algorithm, but is useful in its analysis.
The vertices whose degree is one are called *hanging* vertices. Each vertex eventually becomes hanging. Observe that level(v) is the value assigned in the algorithm to v (in other words, it is the number of times we perform operation Halve until v becomes hanging).

Each vertex v becomes a hanging vertex in the statement (#); its only neighbor in this moment is some vertex w. Denote father(v) = w. The vertex w is called a *nonleaf* vertex if it is a *father* of some vertex v; in this case v is called a *son* of w.

Denote by Sons(w) the set of all vertices which become sons of w in the course of the algorithm.

Claim B
For each nonleaf node w the following inequality holds:

$$(\#\#) \quad \deg_G(w) \geq \frac{1}{2} \sum_{v \in \mathsf{Sons}(w)} \deg_G(v).$$

Proof (of the claim). Let $d_i(w)$ be the degree of w just before the $(i+1)$-th application of Halve, $k_i = |\{u \in \text{Sons}(w) \ : \ \text{level}(v) = i \text{ and } l = \text{level}(w)\}|$. Then for each $0 \leq i < l$

$$d_{i+1} \leq (d_i(w) - k_i)/2 + \tfrac{1}{2}.$$

This easily implies the following inequality:

$$d_0 \geq \textstyle\sum_{i=0}^{l-1} 2^i k_i = \sum_{v \in \mathsf{Sons}(w)} 2^{\text{level}(v)}.$$

At the same time for each node v we have $2^{\text{level}(v)+1} \geq \deg(v)$, since only level$(v)$ *halvings* are enough to make v a hanging vertex. Hence

$$2^{\text{level}(v)} \geq \tfrac{1}{2} \cdot \deg(v).$$

Observe that $d_0(w) = \deg_G(w)$, where G is an initial graph. The inequalities above imply directly (##).

□

The thesis now follows from the fact that all edges incident to nonleaf vertices are contained in Incident(M), owing to Claim A. The condition (##) in Claim B says that $|\,\text{Incident}(M')| \geq \frac{1}{2}|E - \text{Incident}(M')|$. This directly implies $|\,\text{Incident}(M')| \geq |E|/3$.

□

Each graph can be easily decomposed into $\log n$ bipartite graphs; the k-th such graph consists of all edges joining vertices v_i, v_j such that k is the first binary digit from the right distinguishing i and j (if they are written in binary). Then we can apply our parallel matching algorithm to each of such bipartite graphs successively. This proves the following theorem.

Theorem 7.1.4
An inclusion maximal matching can be found in $\log^4 n$ *time with* $O(n+m)$ *processors of an* EREW PRAM.

Remark (about operation *Halve* in general graphs). The operation Halve for general (nonbipartite graphs) is similar. However, we first have to construct an Euler cycle of the graph, and refer the reader to [GR88]. A special vertex s is added and all odd-degree vertices are connected to s. Then the graph has an Euler cycle (if connected). This cycle is used in the operation Halve.

7.2 An application of randomized handshaking

The randomized algorithm for maximal matchings is much simpler than the deterministic one. The main advantage of the randomization here is to gain simplicity.

Recall that in Chapter 2 we introduced two types of randomized handshaking: one-phase handshaking and two-phase handshaking. We have seen (in Chapter 4) in an example of a complete graph that they can produce a large matching after a single application, though the matching produced is not necessarily inclusion maximal. The overall structure of the randomized algorithm is to apply two types of handshaking until the graph becomes empty.

```
Algorithm Randomized_Maximal_Match;
M := ∅;
while E ≠ ∅ do
    begin
    X := Randomized_Partial_Match(G)
    M := M ∪ X;
    E:= E − Incident(X);
    end;
return M; {M is a maximal matching}
```

The function Randomized_Partial_Match is realized by a single application of two-phase and one-phase handshaking as described in Chapter 4.

```
function Randomized_Partial_Match(G);
{returns a partial matching M'}
Step 1.    Apply two-phase randomized handshaking to G;
Step 2.    Apply one-phase randomized handshaking to G;
Step 3.    M' := set of edges corresponding
           to mutual handshakes done in Step 2;
   return M';
```

The function returns a large matching for a complete graph (as we saw in Chapter 2). However, this is not true in general. For example the expected size of Randomized_Partial_Match($K_{3,100}$) is very close to 3. So we choose with high probability only two edges, but we remove at least 200 edges. Hence the effectiveness of the function Randomized_Partial_Match(G) is that it implies the removal of a large number of edges: on average at least $|E|/8$ edges are removed. In fact the example of $K_{3,100}$ shows that many more edges are sometimes removed.

Lemma 7.2.1
Let $M =$ Randomized_Partial_Match(G), *where* $G = (V, E)$. *Then the average cardinality of* Incident(M) *is at least* $|E|/8$.

Proof We introduced the notion of a *φ-good* vertex in Chapter 2. Recall that v is φ-good iff $\varphi(w) \leq \varphi(v)$ for at least $\lceil \deg(v)/3 \rceil$ neighbors w of v. Recall also that an edge was defined to be φ-good iff at least one of its endpoints is φ-good.

Let $\varphi(v) = \deg(v)$ for each vertex v. We say (to simplify notation) that a vertex is *good* iff it is φ-good. Similarly the edge is *good* if one of its endpoints is good.

Claim Let M = Randomized_Partial_Match(G) and let v be a good vertex. Then

$$Prob\{v \text{ is matched by } M\} \geq 1/4.$$

Let $u_1, u_2, \ldots, u_k$ be neighbors of v such that $\deg(w) \leq \deg(v)$. From the assumption that v is good we have

$$k \geq \deg(v)/3 \text{ and } \deg(u_i) \leq 3k \text{ for } 1 \leq i \leq k.$$

The probability that none of the vertices u_i *gives a hand* to v in Step 1 (in the function Randomized_Partial_Match) is

$$\left(1 - \frac{1}{\deg(u_1)}\right) \cdot \left(1 - \frac{1}{\deg(u_2)}\right) \cdot \ldots \cdot \left(1 - \frac{1}{\deg(u_k)}\right)$$
$$\leq \left(1 - \frac{1}{3k}\right)^k = \left(1 - \frac{1}{3k}\right)^{3k \cdot \frac{1}{3}} \leq e^{-1/3} < \frac{1}{2}.$$

Hence the probability that some of the neighbors of v *gives a hand* to v in Step 1 is at least 1/2, which means that after Step 1 the degree of v is at least 1. If v has a nonzero degree then the probability that none of its incident edges is not a handshake in Step 2 is at most 1/2. All together the probability that v is incident to a returned matching is at least $1/2 \cdot 1/2 = 1/4$. This completes the proof of the claim.

Assume e is a good edge. Then one of its endpoints v is a good vertex and with a probability of at least 1/4 is matched by a partial match computed in one iteration. However, if v is matched then the edge e is removed before the next iteration starts. Hence each good edge is removed with probability at least 1/4.

There are at least $|E|/2$ good edges, owing to Lemma 2.6.1 from Chapter 2. Hence on average at least $|E|/8$ edges are removed if E is the set of edges at the beginning of a given iteration.

□

Now Lemma 7.2.1 directly implies the following result.

Theorem 7.2.2 ([II86])
The expected number of iterations of the algorithm Randomized_Maximal_Match *is logarithmic. An inclusion maximal matching can be computed in randomized* $O(\log n)$ *time with* $O(n+m)$ *processors on a* CRCW PRAM.

7.3 Bibliographic notes

The randomized algorithm for maximal matching is from Israeli and Itai [II86] and the deterministic algorithm is from Kelsen [K94].

8

Maximal independent sets

(Inclusion) maximal matchings are directly related to (inclusion) maximal independent sets. A set X of vertices is independent in a given graph iff no two different elements of X are neighbors in G.

The *line graph* of G is a graph L whose vertices are edges of G. Two vertices are neighbors in L if they have a common endpoint (as edges) in G. Then a set M of edges of G is a matching in G iff it is independent (as a set of vertices) in L.

The NC-construction of a maximal independent set is also related to matchings in the following way: the basic part of the algorithm is a parallel construction of a matching with *sufficiently small* weight.

In this chapter we show how to construct in parallel a maximal independent set. Of course construction of a maximum cardinality independent set is a much harder problem, since it is *NP-complete.*

We present two very simple randomized algorithms due to Luby [L86], one of which can be derandomized using the technique from Chapter 4. However, the number of processors of a derandomized version is too large. We then show a completely different and more efficient NC-algorithm, due to Goldberg and Spencer.

We also mention some extensions to *hypergraphs.* The maximal independent sets problem is much harder for hypergraphs than for standard graphs and no NC-algorithm is known to solve it. Another generalisation of maximal independent sets is the notion of a k-dependent set, we cover this subject at the end of the chapter.

8.1 Two algorithms of Luby

The first algorithm of Luby [L86] is extremely simple, but the second has the advantage that it can be derandomized. Both of them are based on a fact presented below. Recall that Neigh(X) is the set of all neighbors of vertices in X.

Remark We can randomly permute the vertices by randomly assigning to them integers from the interval $[1..n^4]$. The probability that two numbers are the same is very small, and we obtain the required random renumbering number(*).

```
Permutation algorithm ;
IndepSet := ∅;
while E ≠ ∅ do
  begin
   permute randomly all vertices; {the vertices are renumbered}
   let number(*) be the corresponding numbering;
   for each v ∈ V do in parallel
      if number(v) < number(w) for each w ∈ Neigh(v);
         then mark v;
   let X be the set of marked vertices;
   IndepSet := IndepSet ∪X;
   CLEAN-UP(G, X)
  end
return IndepSet;
```

where the operation CLEAN-UP is defined as follows:

```
procedure CLEAN-UP(G, X)
remove all edges incident to X ∪ Neigh(X); remove from G all vertices
(X ∪ Neigh(X));
```

Lemma 8.1.1 *(key lemma)*
Assume that Prob{v is marked} $\geq c/\deg(v)$, *for each vertex* $v \in V$, *where* $\frac{1}{4} \leq c \leq 1$ *is a constant. Assume also that the events "i-th vertex is marked" are pairwise independent. Let X be the set of marked vertices.*

Then the expected number of edges incident to $\text{Neigh}(X) \cup X$ *is at least* $|E|/36$.

Proof We say here that v is *good* iff $\deg(w) \leq \deg(v)$ for at least $\lceil \deg(v)/3 \rceil$ neighbors w of v.

Claim
If a vertex v is good then the probability that v is a neighbor of a marked vertex is at least 1/18.

Assume $k = \deg(v)$ and $v_1, v_2, \ldots, v_k$ are neighbors of v. Introduce random variables: $x_i = 1$ iff v is marked. If v is good then the sum of all the x_i is at least 1/12 since there are at least k/3 neighbors marked with probability at least c/k and $\frac{1}{4} \leq c$.

The vertex v is a neighbor of a marked vertex if at least one of variables x_i equals 1. Now the claim follows from Theorem 4.5.1 which implies that the corresponding probability is at least $\frac{2}{3} \cdot \frac{1}{12} = \frac{1}{18}$. There are at least $|E|/2$ edges incident to good vertices, owing to Lemma 2.6.1 from Chapter 2. Each one of its edges is incident to a vertex in $X \cup \text{Neigh}(X)$ with probability at least $\frac{1}{18}$.

Hence the expected number of edges which have this property is at least $\frac{1}{36} \cdot |E|$.

□

Theorem 8.1.2
The permutation algorithm finds a maximal independent set in expected time $O(\log n)$ *with* $O(n+m)$ *processors.*

Proof The probability that a given vertex is marked is $1/k$, where $k = \deg(v)$. This fact follows directly from the following claim.

Claim Assume we randomly permute k elements. The probability that a fixed element is on the first position is $1/k$.

There are k permutations, and exactly $(k-1)$ permutations with the fixed element in the first position. This proves the claim.

The expected number of removed edges in each connected component of the graph is a proportion of the set of all edges with coefficient $1/36$, from Lemma 8.1.1. Now we can apply Lemma 4.6.1 from Chapter 4, which implies that the expected number of iterations is logarithmic.

□

For each vertex v we define the set of its *large neighbors*:

$$\text{LargeNeigh}(v) = \{w \in \text{Neigh}(v) \colon \deg(w) \geq \deg(v)\}.$$

Theorem 8.1.3
Assume the events in the MARKING *step of the algorithm are pairwise independent. Then the algorithm* Vertex-Degree *finds a maximal independent set in expected time* $O(\log n)$.

Proof It is enough to show the following claim. The rest is the same as in the proof of Theorem 8.1.2.

Claim For a fixed vertex v the probability that v is marked immediately after the MARKING and UNMARKING steps is at least $1/(4 \cdot \deg(v))$.

Observe that for any other vertex w the probability that w was marked does not depend on whether v was marked in the step MARKING, because of pairwise independence. Assume that v was marked; then the probability that v is later unmarked equals the probability that any element of LargeNeigh(v) was marked. Observe that

$$|\,\text{LargeNeigh}(v)| \leq \deg(v).$$

Then *Prob*{an element of LargeNeigh(v) was marked} is not larger than

$$\textstyle\sum_{w \in \text{LargeNeigh}(v)} 1/(2 \cdot \deg(w) \leq \sum_{w \in \text{LargeNeigh}(v)} 1/(2 \cdot \deg(v)) \;=\; 1/2.$$

Hence the vertex is marked in the MARKING step with probability $1/(2 \cdot \deg(v))$ and unmarked in the UNMARKING step with probability at most $\frac{1}{2}$. Hence the final probability is at least $1/(4 \cdot \deg(v))$.

□

```
Algorithm Vertex-Degree;
IndepSet := ∅;
while E ≠ ∅ do
  begin
    MARKING:
      for each v ∈ V do in parallel
        mark v with probability 1/(2 · deg(v));
    UNMARKING:
      for each v ∈ V do in parallel
        if LargeNeigh(v) contains a marked vertex then unmark v;
    let X be the set of marked vertices;
    IndepSet := IndepSet ∪X;
    CLEAN-UP(G, X)
  end
return IndepSet;
```

Theorem 8.1.3 and the derandomization theorem from Chapter 4 directly imply the following result. Observe that the algorithm mentioned in the theorem below uses at least a quadratic number of processors, so it is much less efficient then randomized ones (but it is deterministic!). In the next section we present more efficient deterministic algorithms based on a combinatorial approach.

Theorem 8.1.4
There exists an NC-algorithm for the construction of a maximal independent set.

8.2 Deterministic algorithm for maximal independent sets

In this section we present a version of the *Goldberg and Spencer algorithm* for maximal matchings. First we describe the procedure Tough_Indep_Set which finds a *tough* independent set X. We say that X is *tough* iff the following inequality holds:

$$|X \cup \text{Neigh}(X)| \geq \tfrac{1}{4}n/\log(n).$$

The procedure Tough_Indep_Set maintains a family $\mathcal{F}$ of disjoint independent sets, with initially $\mathcal{F}$ consisting of singletons (each vertex is a singleton).

The main part is the operation PAIRING which merges disjoint pairs of independent sets. After executing the operation PAIRING the cardinality of $\mathcal{F}$ is halved, but in the same time components of $\mathcal{F}$ grow larger. At the end we have a tough independent set.

The core of the operation PAIRING is the removal of sets of *conflicting* vertices. The number of removed elements is minimized by reducing this problem to the computation of a *relatively small* weighted matching in a weighted graph. This is done by using a very rough algorithm which employs the following combinatorial result.

Lemma 8.2.1
Each complete graph K_{2k} can be decomposed into $2k-1$ pairwise disjoint perfect matchings. For each edge we can compute in $O(1)$ time with one processor the matchings to which this edge belongs.

Proof Denote n as $2k$. Assume the set of vertices V is $[0 \ldots n-1]$. The matchings correspond to edge colorings of K_{2k} with colors $0 \ldots n-2$.

The color of the edge (i,j) is given by the following formula:

$$\text{if } i,j < n-1 \text{ then } \text{color}(i,j) = (i+j) \bmod (n-1).$$
$$\text{if } i < n-1 \text{ and } j = n-1 \text{ then } \text{color}(i,j) = 2i \bmod (n-1).$$

□

Figure 8.1 shows the decomposition of K_6 by applying the above formula. The colors of an edge give the index of a perfect matching to which this edge belongs.

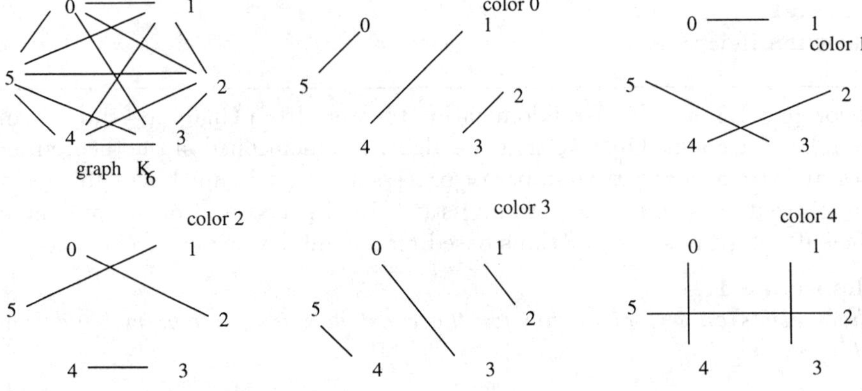

Fig. 8.1 Edge coloring of edges of K_6 with five colors $0 \ldots 4$.

Assume $G = (V, E)$ is a weighted undirected graph with an even number n of vertices. Let $V = [0 \ldots n-1]$. Denote by SmallPerfMatch(G) any perfect matching of K_n. Then

$$M = \{(i_1, j_1), (i_2, j_2), \ldots, (i_{n/2}, j_{n/2})\}$$

such that

$$\sum_{e \in M \cap E} \text{weight}(e) \leq \Big(\sum_{e \in E} \text{weight}(e)\Big)/(n-1).$$

Lemma 8.2.2
Assume $G = (V, E)$ is a weighted undirected graph with an even number n of vertices and with m edges. Let $V = [0 \ldots n-1]$.

Then we can compute SmallPerfMatch(G) *in* $\log n$ *time with* $O(n+m)$ *processors.*

Proof Owing to Lemma 8.2.1 we can color the edges of G; each edge gets a color as if it was in K_n. However, only m edges of G are considered. Then we have a collection of $n-1$ disjoint sets of edges. It is easy to compute the one of minimum total weight in $O(\log n)$ time with $O(n+m)$ processors. It corresponds to some color i. Then we take as M the perfect matching in K_n implied by color i, according to decomposition from Lemma 8.2.1.

□

For two disjoint subsets $C_1, C_2 \subseteq V$ denote

$$\text{ConflictSet}(C_1, C_2) = (C_1 \cap \text{Neigh}(C_2)) \cup (C_2 \cap \text{Neigh}(C_1)).$$

Let $\mathcal{F} = (C_1, C_2, \ldots, C_k)$ be a collection of disjoint subsets of V. Denote by ConflictGraph$(\mathcal{F})$ the graph K_k with weights

$$\text{weight}(i,j) = |\,\text{ConflictSet}(C_i, C_j)|.$$

Define the operation

$$\text{MERGE}(C_1, C_2) = C_1 \cup C_2 - \text{ConflictSet}(C_1, C_2).$$

We describe below the algorithm computing a tough independent subset.

```
function Tough_Indep_Set (G);
   {assume w.l.o.g. that n is a power of two}
k := n;
F := ({v1}, {v2}, ..., {vn});
while F contains no tough subset do
   begin
     K := ConflictGraph(F);
     {(i1, j1), (i2, j2), ..., (ik/2, jk/2)} := SmallPerfMatch(K);
   PAIRING: for each 1 ≤ t ≤ k/2 do in parallel
                  Ct := MERGE(C_it, C_jt);
     k := k/2; F := (C1, C2, ... Ck)
   end
return any tough subset Ci ∈ F.
```

The whole NC-algorithm for a maximal independent set problem is now very simple.

```
Algorithm DETERMINISTIC-MIS;
IndepSet := ∅;
while E ≠ ∅ do
  begin
    X := Tough_Indep_Set(G)
    IndepSet := IndepSet ∪X;
    CLEAN-UP(G, X)
  end
return IndepSet;
```

Theorem 8.2.3
The algorithm DETERMINISTIC-MIS *will find a maximal independent set in* $O(\log^4 n)$ *time with* $O(n+m)$ *processors.*

Proof We show three claims which directly imply the theorem.

Claim 1.
Assume none of the sets in $\mathcal{F}$ is tough. Then in the single operation PAIRING, in total of at most $\frac{1}{2}n/\log(n)$ vertices are removed in all operations MERGE executed in this instance of PAIRING.

Proof (of the claim) Let $K :=$ ConflictGraph$(\mathcal{F})$. It is easy to see the following fact:

if $\mathcal{F}$ contains no tough set then the total weight of K is at most $k \cdot \frac{1}{4}n/\log(n)$.

This implies by Lemma 8.2.2 that SmallPerfMatch(K) has weight at most $(k/(k-1)) \cdot \frac{1}{4}n/\log(n)$. This fact and inequality $k/(k-1) \leq 2$ directly imply the claim.

□

Claim 2.
Assume $n \geq 4$. Then the function Tough_Indep_Set returns a tough independent set.

Proof Assume the function is incorrect: no tough set ever appears in $\mathcal{F}$. Then after $\log n$ iterations there is only one set C_1 in $\mathcal{F}$. If no vertices were removed then C_1 would have n elements. However, in total at most $\log(n) \cdot \frac{1}{2}n/\log(n)$ elements are removed in $\log n$ instances of PAIRING. Hence at most $n/2$ elements are removed and we have an independent set with at least $n/2$ elements. This set is tough, so the function is correct.

□

Claim 3.
The function Tough_Indep_Set is iterated $O(\log^2)$ times in the algorithm DETERMINISTIC-MIS.

Operation CLEAN-UP removes part of the vertices from V after each iteration. In each iteration we remove at least $\frac{1}{4}k/\log(k)$, if the actual number of vertices in V is k. Initially $|V| = n$. It is easy to see that after $O(\log^2 n)$ iterations no vertices remain. This completes the proof of the claim and (also) of the theorem.

□

8.3 Maximal independent sets in hypergraphs

At the end of this chapter we mention some extensions to hypergraphs. A *d-dimensional hypergraph* $H = (V, E)$, where E is a family of subsets of V, each one of cardinality at most d. If $d = 2$ then the elements of E are two-element sets corresponding to the usual edges. In general the elements of E are called *hyperedges*. A set $X \subseteq V$ is *independent in the hypergraph H* iff for no hyperedge $e \in E$ we have $e \subseteq X$. X is maximal iff it is not a proper subset of another independent set.

Dahlhaus, Karpinski and Kelsen extended Theorem 8.2.3 to hypergraphs of dimension 3. The algorithm DETERMINISTIC-MIS is easily extendible to work for such hypergraphs, and we refer to [DKK92].

We redefine here the definition of Neigh(X). Let

$$\text{Neigh}(X) = \{v \in V \;:\; \text{there is a hyperedge } e \text{ such that } e \subseteq X \cup \{v\}\}.$$

If $d = 3$ then we can redefine the concepts ConflictSet(C_1, C_2), MERGE(C_1, C_2), and ConflictGraph$(\mathcal{F})$ using the new definition of Neigh. Then the same algorithm (as in the preceding section) works since the following basic fact still holds:

Fact 8.3.1
If C_1, C_2 are independent sets in a 3-hypergraph then $C_1 \cup C_2 - \text{ConflictSet}(C_1, C_2)$ is also an independent set.

Theorem 8.3.2
A maximal independent set in a hypergraph of dimension 3 can be found in $O(\log^4 n)$ time with $O(n + m)$ processors.

Proof Use the algorithm DETERMINISTIC-MIS with the new definition of Neigh.

□

The extensions to hypergraphs of larger dimensions is quite complicated, since Fact 8.3.1 does not generally hold for such hypergraphs. However, if d is a constant then it was shown in [Ke92] that there is an *RNC*-algorithm for maximal f-matching in d-hypergraphs.

First we generalize the concept of the *neighbor-set*. Let

$$\text{Neigh}_k(X) = \{Y \subseteq V \;:\; \text{there is a hyperedge } e \text{ such that } e \subseteq X \cup Y \text{ and } |Y| = k \text{ and } X \cap Y = \emptyset\}.$$

We also redefine the maximum degree of the graph. Let

$$\Delta_k(H) = \max\{|\,\mathrm{Neigh}_k(X)|^{1/k} \;:\; X \subseteq V\}$$
$$\Delta(H) = \max\{\,\Delta_k(H)\}.$$

The correct, complicated, proof of the theorem below was shown by Kelsen in [Ke92], but is beyond the scope of this text.

Theorem 8.3.3 (CiteKe92)
Assume the dimension d of the hypergraph is $O(1)$. Then the Beame–Luby algorithm terminates after $(\log(n))^{O(1)}$ iterations, so the maximal independent set problem for hypergraphs of constant dimension is in RNC.

```
Beame–Luby Algorithm;
IndepSet := ∅;
while E ≠ ∅ do
   begin
     MARKING:
        for each v ∈ V do in parallel
           mark v with probability 1/(2^(d+1) · Δ(H));
           mark all isolated vertices;
     UNMARKING:
        for each e ∈ E do in parallel
           if all nodes in e are marked
              then unmark all of them;
     let X be the set of marked vertices;
     IndepSet := IndepSet ∪X ;
     V := V − X; E := {e ∩ V  : e ∈ E };
     V := V − Neigh_1(X);
     unmark all nodes;
   end
return IndepSet;
```

8.4 Maximal k-dependent sets

The notion of independent set can be naturally generalized by weakening the requirement on vertex nonadjacency or edge no-incidency.

A subset of the set of vertices of a graph is k-dependent if no vertex in the subset is adjacent to more than k vertices in the subset. Note that a 0-dependent set in G is an independent set of vertices of G. A 1-dependent set is in general a set of independent vertices and edges, while a 2-dependent set is a set of independent paths (possibly degenerate) and cycles such that no two nonconsecutive vertices of these paths or cycles are adjacent. The cardinality k-dependent set is *NP*-hard for each fixed k and the problem of finding a maximal k-dependent set can be solved by trivial greedy algorithms in linear time.

For a given subgraph H of G, and for any vertex v of G, define the degree of v in H as the number of neighbors of v which are vertices of the subgraph H, and denote it as $d(H,v)$. Note that in the last definition the vertex v is not necessarily in H. If $S \subseteq V$ then $\gamma(S)$ will denote the subgraph induced by S. This subgraph has vertex set S, and its edge set $E_{\gamma(S)}$ consists of these edges in E that are incident only to vertices in S.

The parallel algorithm for a maximal k-dependent ($k > 0$) set can be seen as a reduction of the problem of constructing a maximal 0-dependent (independent) set.

Algorithm Mds(G);
$\{$returns a maximal k-dependent set Q for $G = (V, E)\}$
  $Q \leftarrow$ Maximal-Independent-Set(G);
  $R \leftarrow V \setminus Q$;
  $B \leftarrow \{v \in R \mid d(\gamma(Q), v) \leq k\}$;
  while $B \neq \emptyset$ **do**
    $H \leftarrow$ the graph whose set of vertices is B such that (v, w) is an edge of H iff v and w have a common neighbor in Q or (v, w) is also an edge in G;
  $M \leftarrow$ Maximal-Independent-Set(H);
  $Q \leftarrow Q \cup M$;
  $S \leftarrow \{v \in Q \mid d(\gamma(Q), v) = k\}$;
  $R \leftarrow R \setminus (M \cup \text{Neigh}_{\gamma(R)}(S))$;
  $B \leftarrow \{v \in R \mid d(\gamma(Q), v) \leq k\}$;
  output Q;

Lemma 8.4.1
Algorithm Mds *is partially correct, that is if it stops then the set Q to output is a maximal k-dependent set in G.*

Proof It is sufficient to observe that the augmentation of Q by M is correct since M is in particular independent in G, no two vertices in M share a common neighbor in Q and no vertex in M is a neighbor of a vertex in Q that already has k neighbors in Q.

□

Lemma 8.4.2
The block under the **while** *statement is iterated $\mathcal{O}(k^2)$ times.*

Proof For a vertex $v \in G$, let $\text{cap}(v) = \min(d(G,v), k) - |\text{Neigh}_{\gamma(Q)}(v)|$ at a given stage of performance of Algorithm Mds. Consider a vertex $v \in B$ at the beginning of the i-th iteration of the block. Next, let

$$g(v) = \sum_{w \in \text{Neigh}_{\gamma(Q)}(v)} \text{cap}(w).$$

Note that $g(v)$ is always bounded by k^2 from above.

Also, if v disappears from B in some of the next iterations then it can never reappear in B. On the other hand, after each iteration, if v remains in B then there exists a vertex w newly inserted into Q that either shares a neighbor in Q with v or is itself a neighbor of v in G. In the first case, $g(v)$ decreases at least by 1. In the second case $g(v)$ increases by $\text{cap}(w)$, that is at most by k. However, the second case can occur at most k times since $\text{cap}(v)$ cannot be negative if v is to stay in B.

Since $\text{cap}(w)$ for each neighbor w of v in Q has to be positive in order to keep v in B, we conclude that after $k^2 + k$ iterations v has to disappear from B.

Suppose that v is not a member of the original set B. This means that v has no neighbor in the original set Q. Therefore, it could be added to the original set Q preserving its independence property which would contradict the maximality of the set.

We obtain a contradiction. Thus, we can conclude that all vertices that appear in the sets B are members of the original set B. Combining this conclusion with the fact that no member of B can survive more than the $k^2 + k$ iterations we obtain the thesis of the lemma.

□

Combining the above two lemmas, we obtain the correctness of Algorithm Mds.

Theorem 8.4.3
Algorithm Mds *is correct.*

Lemma 8.4.4
Suppose that a maximal independent set in a graph on n nodes can be found in time $T(n)$ with $P(n)$ processors. Algorithm Mds *can be implemented in time $\mathcal{O}(\log n + T(n))$ using a* PRAM *with $\mathcal{O}(n^2 + P(n))$ processors.*

Proof A maximal independent set in G as well as a maximal independent set in the auxiliary graph H can be found in time $T(n)$ using $P(n)$ processors. By Lemma 2, we can replace the **while** statement by a **for** loop with the number of iterations $\mathcal{O}(k^2)$, in this way avoiding the test for the emptiness of B.

We can implement the set operations in constant time by representing all the involved sets B, Q, R, S with n-element vectors, each of them with 1 in the i-th position if and only if the i-th vertex in G is currently in the set.

The construction of the auxiliary graph H, in particular finding all pairs (v, w) such that v and w have a common neighbor in Q, seems to be more costly. It immediately reduces to the following problem: given a bipartite graph $F = (V_1, V_2, E)$, where the degree of each vertex in V_2 is bounded by k, construct the graph $F' = (V_2, E')$ such that (v, w) is in E' if and only if v and w have a common neighbor in V_1. Suppose that $V_2 = \{1, 2, \ldots, s\}$.

Construct a matrix of size $s \times s$ such that $W(i, j)$, $1 \le i \le j \le s$, is set to the list of neighbors of i in F (i.e. in V_1). Note that such a list can have at most k elements.

Now the construction of the graph F' becomes easy. For each pair i, j where $1 \leq i \leq j \leq s$, we check whether the lists $W(i,j)$ and $W(j,i)$ have at least one element in common. If so we augment E' by (i,j), that is we set to 1 the corresponding entry of the adjacency matrix of F'. Note that comparing two such lists takes time $\mathcal{O}(k)$. We conclude that F' can be constructed in time $\mathcal{O}(\log s)$ using $\mathcal{O}(s^2)$ processors.

Consequently, the auxiliary graph H can be constructed in time $\mathcal{O}(\log n)$ using $\mathcal{O}(n^2)$ processors. All instructions within the loop are executed only $\mathcal{O}(k^2)$ times, and we conclude that Algorithm Mds can be implemented in time $\mathcal{O}(\log n + T(n))$ using $\mathcal{O}(n^2 + P(n))$ processors.

□

The following result follows from the last lemmas.

Theorem 8.4.5
Let k be a nonnegative integer. A maximal k-dependent set in a graph on n nodes can be computed in time $\mathcal{O}(\log^4 n)$ using $\mathcal{O}(n^2)$ processors.

8.5 Bibliographic notes

The randomized algorithms for maximal independent sets are from Luby [L86]. The deterministic algorithm is from Goldberg and Spencer [GS89]. The algorithms for hypergraphs are from Dahlhaus, Karpinski and Kelsen [DKK92] and Kelsen [Ke92]. The algorithm for maximal k-dependent sets is from Diks, Garrido and Lingas [DGL93].

9

Four easy subclasses of graphs

In this chapter we consider three easy cases of matchings in special graphs: trees, dense graphs and regular bipartite graphs. They are *easy* in the sense of possessing simple NC-algorithms for the matching problem.

Usually efficient sequential algorithms for trees also have quite efficient parallel implementations. Regular bipartite and dense graphs are among the few known families of graphs which possess perfect matchings.

9.1 Trees

A *tree* is an undirected connected graph without cycles. Assume we choose a vertex called a *root* and direct all edges *top down*. The root is the topmost vertex. The leaves are degree 1 vertices (except the root). A tree can have no perfect matching, even if the number of vertices is even. For example, $K_{1,n}$ has only one edge in the maximum cardinality matching. Our approach to construct an NC-algorithm for trees is to implement a greedy sequential algorithm in parallel.

We define a *leaf-parent* as a vertex such that some of its sons are a *leaf*.

The sequential algorithm proceeds *bottom-up* and is processing leaf-parent nodes. The algorithm is presented below. We implement this algorithm by employing an algebraic approach. For each vertex v which happens to be a leaf-parent in the algorithm SEQ_TREES we set value$(v) = 1$. All other nodes have value equal to 0. We can interpret 0 and 1 as *Boolean* values. Denote by $NAND$ the Boolean operation

$$\text{NAND}(x_1, x_2, \ldots, x_k) = \text{not}(x_1 \wedge x_2 \wedge \ldots \wedge x_k).$$

Theorem 9.1.1
A maximum cardinality matching for a tree of size n can be computed in $O(\log n)$ time with $O(n/\log(n))$ processors on an EREW PRAM.

Proof Assume we placed at each internal node the operation NAND. Then the tree T becomes an expressions tree.

The values of all nodes can be computed using any parallel tree contraction method which works in $O(\log n)$ time with $O(n/\log(n))$ processors, see for example [GR88].

□

The sequential algorithm is presented below.

```
Algorithm SEQ_TREES;
{returns a maximal matching M of a tree T}
M := ∅;
while G ≠ ∅ do
   for each leaf-parent vertex v do
      begin
      choose any leaf w of v; M := M ∪ {(v, w)};
      remove v, w and all edges incident to them;
      end;
return M;
```

The parallel algorithm is an implementation of the algorithm SEQ_TREES and its description is given below. Instead of processing the tree bottom up level by level it uses the tree contraction technique.

```
Algorithm PARALLEL_TREES;
{returns a maximal matching M of a tree T}
M := ∅;
for each leaf v do in parallel
      value(v) := 0;
place at each internal node the operation NAND;
{T becomes an expressions tree}
compute the values of all nodes using any
parallel tree contraction method, see [GR88];
for each vertex v with value(v) = 1 do in parallel
      begin
        choose any son w of v with value(w) = 0;
        insert (v, w) into M;
      end;
return M;
```

9.2 Dense graphs

We say that an undirected n-vertex graph G is *dense* iff $\deg(v) \geq \lfloor n/2 \rfloor$ for each $v \in V$.

Lemma 9.2.1
If the graph is dense then each inclusion maximal matching covers at least $n/2$ vertices.

Proof Assume that an inclusion maximal matching M contains less than $n/2$ vertices. Let v be a vertex not contained in M. Then v has $n/2$ neighbors (at

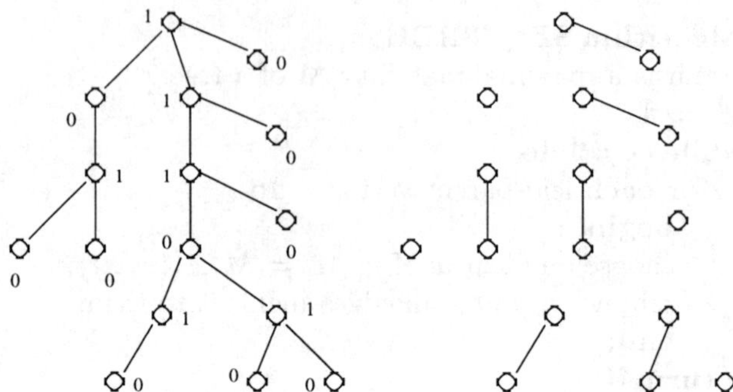

Fig. 9.1 The values computed for each node and a matching produced by the algorithm.

least), and one of them, say w, is not in M. Then none of the endpoints of the edge (v, w) is in M, and we could add this edge to M. This contradicts the maximality of M.

□

Theorem 9.2.2
For each dense graph of an even number n of vertices a perfect matching can be constructed in $\log^4 n$ time with a linear number of processors on a CREW PRAM.

Proof Let us compute any inclusion maximal matching M. Assume the vertices not contained in M can be listed as $v_1, v_2, \ldots, v_{2k}$, where $k \leq n/2$. Figure 9.2 shows a graph with a maximal matching consisting of bold edges. There are four vertices not contained in edges of this matching.

Claim 1
We can construct k disjoint alternating paths $P_1, P_2, \ldots, P_k$ (each one having exactly three edges) such that the path P_i contains vertices v_{2i-1}, v_{2i}. The construction can be done in $\log^4 n$ time with $O(n + m)$ processors.

Such paths P_i are illustrated in Figure 9.3. We augment the matching on each of these paths simultaneously and obtain a perfect matching of the graph. The obtained perfect matching for our example graph is shown on the second graph in Figure 9.3. Hence it is enough to prove the claim.

We construct a bipartite graph G'. Its vertices are pairs (v_{2i-1}, v_{2i}) and all edges of M. The pair v_{2i-1}, v_{2i} and the edge $e_j = (w_1, w_2)$ are connected iff there is an augmenting path consisting of vertices v_{2i-1}, v_{2i} and the edge e_j.

This means that in the original graph G:

there are both edges (v_{2i-1}, w_1), (v_{2i}, w_2), or
there are both edges (v_{2i-1}, w_2), (v_{2i}, w_1).

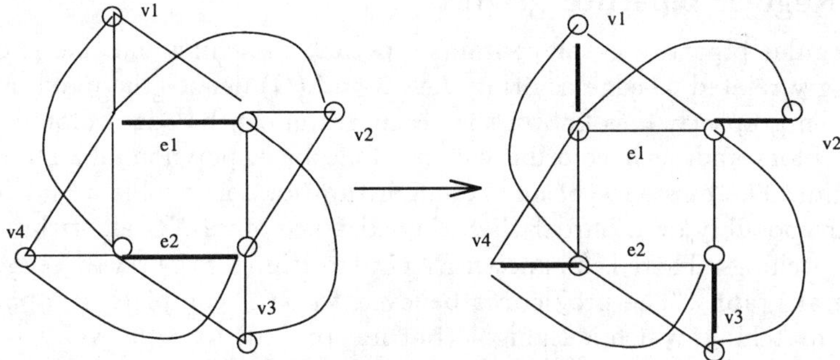

Fig. 9.2 Graph G with an inclusion maximal matching M, and the same graph after changing the matching according to the augmenting paths in the next figure.

The bipartite graph G' for our example graph is shown in Figure 9.3. Let M' be any inclusion maximal matching of G'. Then it is enough to show that each pair v_{2i-1}, v_{2i} is contained in M'. It is obvious that if the degree of each vertex (v_{2i-1}, v_{2i}) is at least k then M' should contain each of the pairs, and we have required number of augmenting pairs. Hence Claim 1 is reduced to the next the claim.

Claim 2
The degree of each vertex (v_{2i-1}, v_{2i}) in the bipartite graph G' is at least k.

Observe that each vertex has degree at least $n/2$ in G and there are no edges between vertices v_i, $1 \leq i \leq 2k$, in the graph G. Then the claim can be proved by a simple calculation, see [DHK93]. This completes the proof of both the claim and the theorem.

□

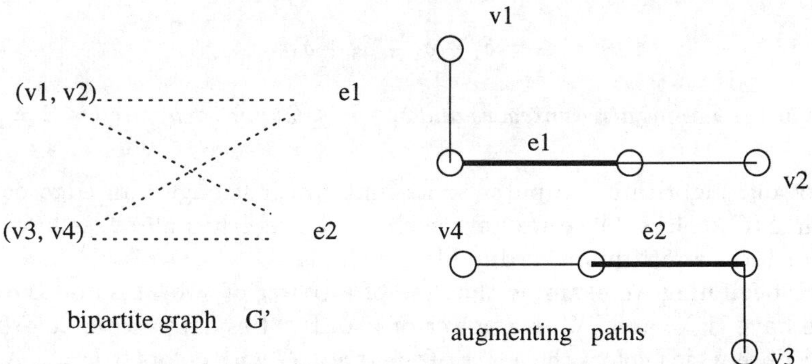

Fig. 9.3 The bipartite graph G' and the augmenting paths in G.

9.3 Regular bipartite graphs

Each regular bipartite graph contains a perfect matching and this phenomenon is strongly related to edge coloring. Let $\Delta = \Delta(G)$ denote the maximum degree of a given graph G. It is known that each graph can be edge colored with only $\Delta + 1$ colors and such coloring can be done by a polynomial time sequential algorithm. The existence of an NC-algorithm for this problem is a hard open problem, possibly even harder than the existence of an NC-algorithm for maximum matchings. There is no known RNC-algorithm for the $(\Delta+1)$-edge-coloring of general graphs. The problem is believed to be P-complete, in opposition to perfect matching, which is thought (but not proved) to be in NC.

The situation is particularly simple for regular bipartite graphs: they can be colored with Δ colors. The edges colored by a specified color form a perfect matching. The coloring can be done by an NC-algorithm. As a side effect of this algorithm the problem of a perfect matching in regular bipartite graphs is shown to be in NC. The corresponding algorithm is optimal up to a logarithmic factor.

It often happens in computer science that people make the following harmless assumption:

assume w.l.o.g. that k is a power of two.

The construction of matchings in regular bipartite graphs is an exception. The assumption that $\Delta(G)$ is a power of two leads to a drastically simpler algorithm, compared with the general case.

However, if we assume that $\Delta(G)$ is not a power of two, then the situation becomes nontrivial. Lev, Pippenger and Valiant have constructed in [LPV81] an elegant algorithm (LPV algorithm) for the general Δ using the following obvious fact.

Lemma 9.3.1
Let 2^k be the largest power of two not exceeding a given natural number $\Delta > 0$. Then there is a decomposition

$$\Delta = \delta_1 + \delta_2 + \delta_3 + \delta_4$$

such that δ_i are nonnegative integers and $\delta_i + \delta_j \leq 2^k$ for each pair $1 \leq i \neq j \leq 4$.

In fact our algorithm computes something much stronger: an edge coloring of G with $\Delta(G)$ colors. Once we have such a coloring, then all edges having the same color form a maximum cardinality matching.

At the beginning we examine the case of a power of two. Assume the set of colors we have is X, and $|X|$ is a power of two. Let Easy_Bipartite_Color(G, X) be a procedure which colors the edges of the graph G with colors from X. Assume that $|X| \geq \Delta(G)$. No two adjacent edges can have the same color. The graph is halved by Halve and each of two "halves" of the graph is colored independently in parallel. This is possible since "halves" of the set X are again powers of two.

```
Algorithm Easy_Bipartite_Color(G, X);
{colors the edges assuming Δ(G) is a power of two}
if |X| = 1 then color G with a single color
{in this case G is a matching}
else while Δ(G) > 1 do
    G1 := Halve(G); G2:=G − G1;
    partition X into two equal sets X1, X2;
    perform in parallel
        Easy_Bipartite_Color(G1, X1)
        and Easy_Bipartite_Color(G1, X1);
```

We use the operation Halve from Chapter 5. The algorithm applies this operation $O(\log n)$ times. The time is $O(\log^2 n)$ on an EREW PRAM. The total work of the algorithm (the product *time* $\cdot$ *processors*) is small, since at each stage we work on a geometrically decreasing graph.

A local edge coloring of the graph G is an assignment of colors to edges local to each vertex. This means that each vertex v colors its side of each of its incident edges, the colors given by this vertex being distinct. However, the edge can have two different colors assigned to it by its endpoints. We call such an edge a *bad edge*. Otherwise the edge here is called *good.*

Observation Assume that G is partially colored, which means that some of the edges are colored, and no two adjacent edges have the same color. The colored edges are colored in the same color by their endpoints. In other words assume we have a partial local coloring Φ, such that all edges which are colored are good. Then we can easily extend the coloring Φ to a full local coloring Φ' in such a way that all edges which were good in Φ are good in Φ'. Denote the operation which is doing that by Locally_Valid_Coloring(G, Φ).

Observation Assume that G is partially colored, and X is a set of good edges. After the operation Locally_Valid_Coloring(G, Φ) each edge of X is still a good edge.

Assume we use the set of colors $\Lambda = [1..\Delta(G)]$. Decompose the set Λ into four sets Λ_i, $i \in [1\ldots4]$, such that the sizes $\delta_i = |\Lambda_i|$ satisfy the condition in Lemma 9.3.1. For each distinct index $i, j \in [1..\Delta(G)]$, denote by $\Phi_{i,j}$ a set of colors which includes $\Lambda_i \cup \Lambda_j$ and such that its cardinality is a power of two. This definition is sound owing to Lemma 9.3.1. If G is locally colored then denote by SubGraph(G, Φ) the subgraph formed by all edges whose both colors (from each endpoint of the edge) are in Φ. Denote by #BadEdges(G) the number of *bad edges* in G.

```
Algorithm Bipartite_EdgeColoring;
{returns maximum matching as a side effect of edge coloring}
Φ := empty coloring;
Locally_Valid_Coloring(G, Φ);
while there is a bad edge in G do
    find i, j which maximize # BadEdges(SubGraph(G, Φ_{i,j}));
    Easy_Bipartite_Color(SubGraph(G, Φ_{i,j}), Φ_{i,j});
    decolor all bad edges in G;
    {Φ becomes a partial coloring}
    Locally_Valid_Coloring(G, Φ);
return all edges colored by 1;
```

Theorem 9.3.2
Assume G is a regular bipartite graph. Then a maximum cardinality matching can be computed in $O(\log^3 n)$ parallel time with $O(n+m)$ processors on an EREW PRAM.

Proof The procedure Easy_Bipartite_Matching works in $O(\log^2 n)$ time since we need $O(\log n)$ iterations and in each of them we use the operation Halve. It is enough to show that the main iteration in the algorithm Bipartite_EdgeColoring is executed a logarithmic number of times.

Observe that there are six sets $\Phi_{i,j}$ for $i \neq j$, $1 \leq i,j \leq 4$. Their union contains all edges, so it also contains all bad edges. Thus some of these sets contain at least $\frac{1}{6}$ bad edges. This directly implies the following claim.

Claim Let i, j be as chosen in a given iteration. Then

$$\#\,\text{BadEdges}(\text{SubGraph}(G, \Phi_{i,j})) \geq \frac{1}{6} \cdot \#\,\text{BadEdges}(G).$$

The claim implies that the number of bad edges decreases logarithmically. Hence the number of main iterations is logarithmic.

□

Example Let us look at the graph illustrated in Figure 9.4. In this case $\Delta = 3$ and we may choose $\delta_1 = 1, \delta_2 = 1, \delta_3 = 1, \delta_4 = 0$. We can take $\Lambda_1 = \{1\}$, $\Lambda_2 = \{2\}$, $\Lambda_3 = \{3\}$.

Assume that the operation Locally_Valid_Coloring(G, Φ) locally colors the edges as shown in the figure. Only edges $(v_1, w_3), (w_1, v_3)$ are good. We choose the subgraph induced by colors {1,2}. It contains three bad edges. After the operation Easy_Bipartite_Color all of them are *good*.

Observe that exactly after this operation the coloring is not locally correct.

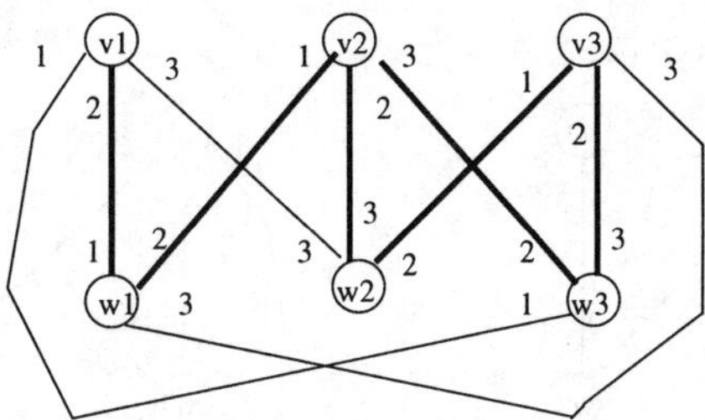

Fig. 9.4 An example bipartite graph after the first LocalColor. There are six bad edges (in bold). There are three bad edges colored with 1 and 2 and these are the first values of i, j which we choose in the algorithm Bipartite_EdgeColoring.

The color 1 appears twice at the vertex w_2. However, all bad edges are decolored and after the next LocalColor we have the situation presented by the second graph in the figure.

In the next iteration we can choose colors $\{2,3\}$.

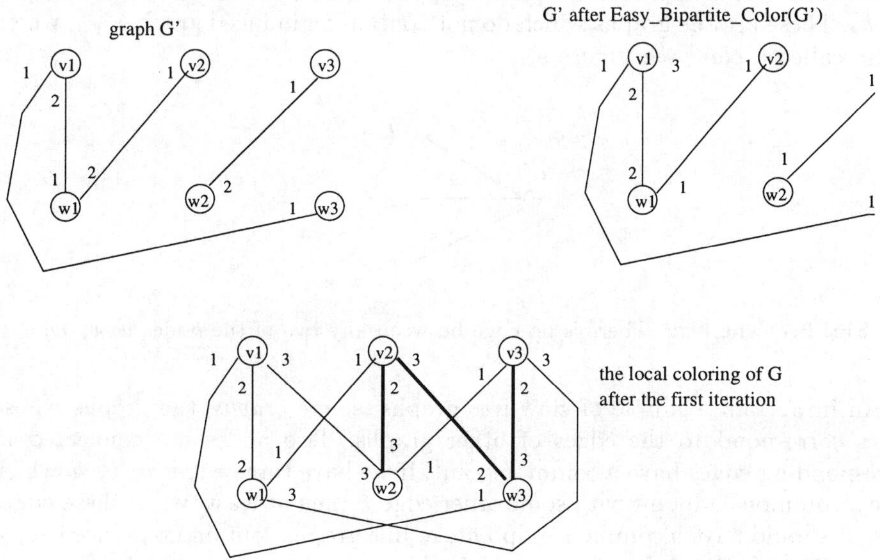

Fig. 9.5 The graph $G' = \text{SubGraph}(G, \{1, 2\})$ before and after the operation Easy_Bipartite_Color(G'). At the bottom, the local coloring of the whole graph; now we have only three bad edges.

All bad edges now belong to SubGraph$(G, Phi_{2,3})$, see Figure 9.6. After this iteration the algorithm terminates. It returns the maximum matching consisting of edges colored by 1: $(v_1, w_3), (w_2, v_3), (w_1, v_2)$.

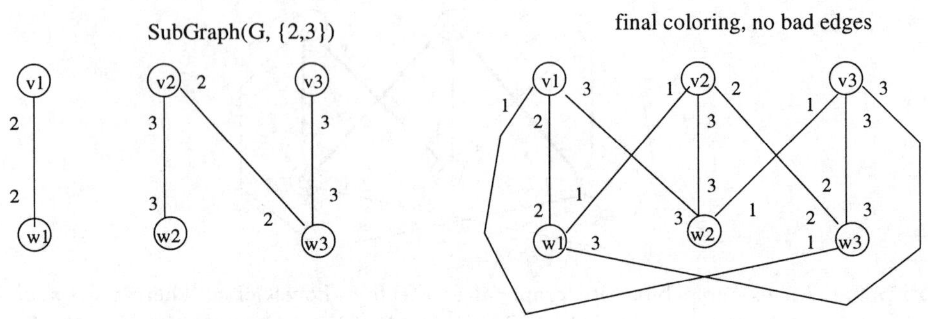

Fig. 9.6 SubGraph$(G, \{2,3\})$ after recoloring and the final edge coloring.

9.4 Claw-free graphs and pseudo-matchings

In this section we show yet another application of NC-computation of *maximal* matchings and also another approach to NC-computation of perfect matching based on the concept of *pseudo-perfect matching.* This again shows "linear algebra at work". This approach works for a special family of graphs called *claw-free graphs.* These are the graphs which do not contain an induced graph $K_{1,3}$, which is also called a *claw*, see Figure 9.7.

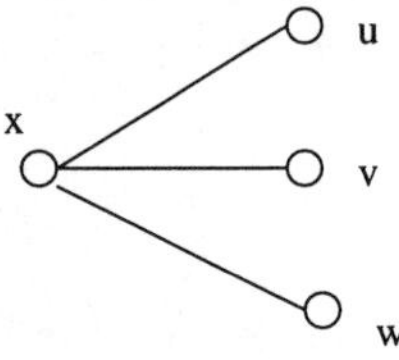

Fig. 9.7 The *claw.* There is no edge between any two of the nodes u, v, w.

An important example of claw-free graphs is *line graphs*, the graphs whose nodes correspond to the edges of other graphs. Two nodes are connected iff corresponding edges have a common point. If we have three edges u, v, w which have a common endpoint with some other edge x then at least two of these edges u, v, w should have a common endpoint, so line graphs contain no induced $K_{1,3}$.

Another family of claw-free graphs is the class of proper interval graphs.

The crucial concept is that of a *pseudo-perfect matching*, a spanning subgraph M of G such that each node has odd degree w.r.t. M.

In this section denote by $\oplus$ the operation of symmetric difference between sets of edges. We can identify this operation with componentwise addition modulo 2 for characteristic vectors corresponding to edge sets. Denote by GF(2) the field {0,1} modulo 2, and by $\mathrm{GF}(2)^m$ the vector space of dimension m over GF(2), where m is the cardinality of the set of edges of G. Each set X of edges can be identified with its characteristic vector which is an element of $\mathrm{GF}(2)^m$.

The set of all cycles of the graph G is a subspace of $\mathrm{GF}(2)^m$ of dimension $n-m+r$, where r is the number of connected components. It is called the *cycle space* of G.

A set of *fundamental cycles* is a collection of cycles composing a basis of this space. Assume G is connected and T is any spanning tree. Then the set of fundamental cycles corresponding to T consists of all cycles formed by adding to T any single non-tree edge.

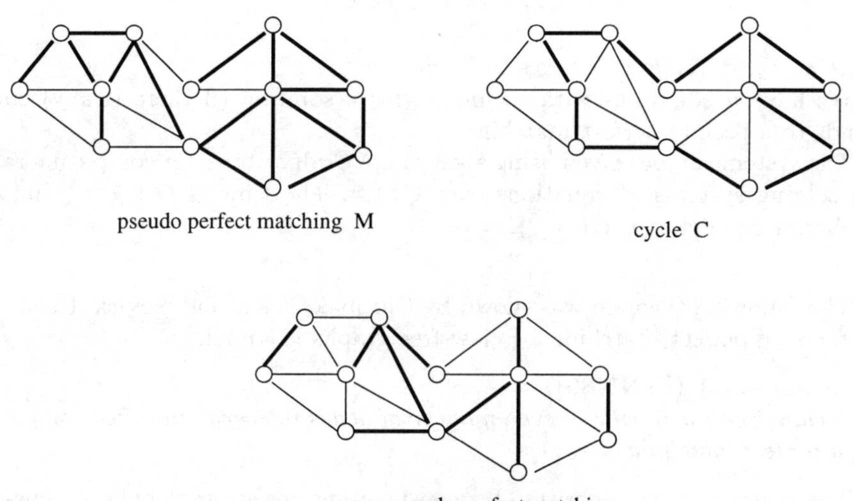

Fig. 9.8 A graph G with a pseudo-perfect matching shown in bold, a cycle C and a better (having less edges) pseudo-perfect matching $M \oplus C$.

An *augmenting cycle* C w.r.t. a pseudo-matching M is a cycle C of G such that $C \oplus M$ is a pseudo-perfect matching having less edges than M. Observe that pseudo-perfect matching of minimal cardinality is a perfect matching. Figure 9.8 shows a pseudo-perfect matching M and an augmenting cycle w.r.t. M.

Augmenting cycles play a similar role as augmenting paths, especially in the case of graphs with maximum degree 3.

Lemma 9.4.1
Let G be an undirected graph with maximum degree 3 having a perfect matching. Let MM be a pseudo-perfect matching. If M is not a perfect matching in G then there exists an augmenting cycle in G.

The most surprising property of pseudo-matchings is that we can easily construct in NC a pseudo-perfect matching (if one exists), despite the fact that it is so closely related to perfect matchings. It is quite possible that an NC-algorithm for perfect matchings will be constructed some time in the future using pseudo-perfect matchings.

Lemma 9.4.2
There is an NC-algorithm to test and construct a pseudo-perfect matching (if there is any).

Proof Let G be a graph with n nodes and m edges. We construct a system of linear equations over GF(2) whose solution is a pseudo-perfect matching. Assign to each edge e a variable $\hat{e}$. For each node v we have an equation

$$\hat{e}_1 \oplus \hat{e}_2 \oplus \ldots \hat{e}_k = 1,$$

where $e_1, e_2, \ldots, e_k$ are all edges incident to v.

We have n equations with m unknowns; a solution (if there is any) corresponds to a pseudo-perfect matching.

The system can be solved using algebraic NC-algorithms for computing ranks and solving systems of equations over GF(2). The time is $O(\log^2 n)$ and the number of processors is $O(m^{5.5})$.

□

The following theorem was shown by Chrobak, Naor, and Novick. Hence the existence of perfect matching for claw-free graphs is trivial.

Theorem 9.4.3 ([CNN89])
Each claw-free graph with an even number of nodes in every connected component has a perfect matching.

The schema of the algorithm for constructing perfect matching in claw-free graphs is as follows. It is based on the following.

Observation If a pseudo-perfect matching of G is an induced forest and G is claw free then M is a perfect matching.

Algorithm Perfect_Matching;
{construct perfect matching for a claw-free graph}
begin
  construct a pseudo-perfect matching M;
  convert M to a pseudo-perfect matching F which is a forest;
  convert F to a pseudo-perfect matching F' which is an induced forest;
  {F' is a perfect matching since G is claw free}
return F';
end algorithm.

The construction of F is quite easy. Assume w.l.o.g. that M consists of a single component. Construct a spanning tree T of M. Let $C_1, C_2, \ldots, C_r$ be the set of fundamental cycles of M with respect to T. Then $M \oplus C_1 \oplus C_2 \oplus \ldots \oplus C_r$ is the required forest F.

The computation of $M \oplus C_1 \oplus C_2 \oplus \ldots \oplus C_r$ can be done effectively as follows. Denote by $P(e,T,M)$ the parity of the number of fundamental cycles w.r.t. T which contain the edge e. For every nontree edge e ($e \notin T$) we have $P(e,T,M) = 1$.

Denote also by $p(v)$ the number of nontree edges incident on v. Let us root the tree T at some node; assume now that T is rooted. Then it is easy to see the following fact.

Fact 9.4.4
Assume $e = (u,v)$, where v is a child of u. Then $P(e,T,M)$ is the sum modulo 2 of $p(w)$ over all w in the subtree rooted at v.

Hence $P(e,T,M)$ can be easily computed in parallel using standard tree contraction techniques, see [GR88]. We can now compute F as the set satisfying

$$F = \{ e\colon e \notin T \text{ or } (e \in T \text{ and } P(e,T,M) = 0) \}.$$

The transformation of F into F' is much harder. We can process each connected component separately in parallel, so assume w.l.o.g. that F is connected, and is a single tree. We say that a node v is a *good cut* of F iff after disconnecting v all hanging subtrees are of size at least $\frac{2}{3}$ times smaller than the whole tree.

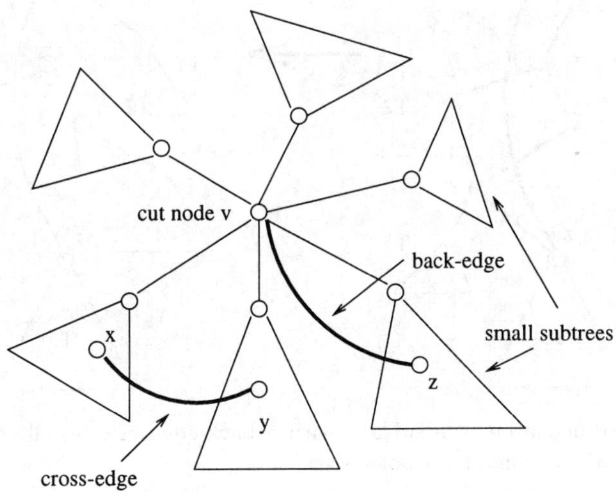

Fig. 9.9 Cross-edges and back-edges w.r.t. v. All hanging subtrees are at least $\frac{2}{3}$ times smaller than the whole tree.

We use a kind of parallel *divide-and-conquer* approach, which consists in parallel processing the connected components of F in such a way that after one iteration the sizes of *bad* components decrease to guarantee that the number of iterations is logarithmic.

We shall describe two stages of the algorithm which reduce the sizes of bad components of F. A connected component of F is a bad component iff it is not an induced tree. The following fact will follow directly from the description of Stage 1 and Stage 2 but we give it in advance to motivate such descriptions.

Lemma 9.4.5 (efficiency of one iteration)
Assume F has a bad component. After performing Stage 1 and Stage 2 the size of the largest bad component of F decreases by at least a factor $\frac{2}{3}$.

Find a node v which is a *good separator* of F and process all subtrees hanging from v in parallel. We can process these subtrees independently without any harm if there are no *bad edges*. These are *cross-edges* (between two distinct hanging subtrees) or *back-edges* (between v and the interior of some subtree), see Figure 9.9.

Stage 1. Elimination of back-edges
Assume (v, z) is a back-edge. Choose z as the lowest node in the subtree with such a property. Then there is a cycle C containing the back-edge and the path from z to v, see Figure 9.10.

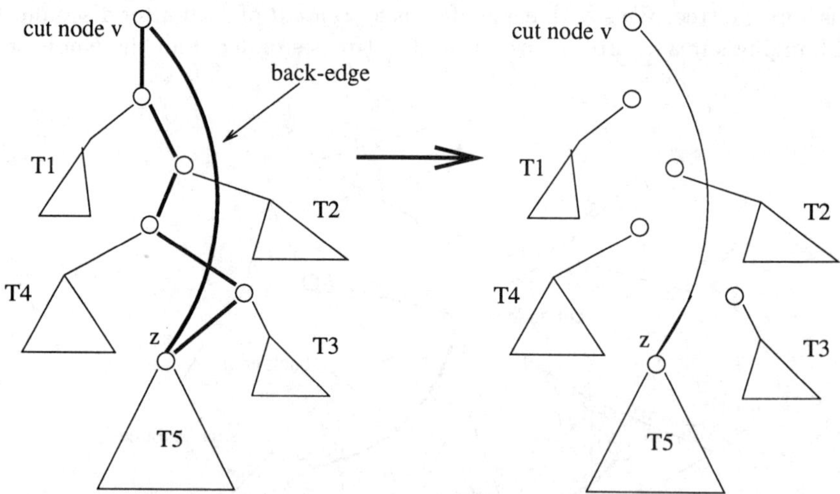

Fig. 9.10 Transformation of a subtree with a back-edge into a collection of smaller subtrees. The cycle C consists of bold edges.

We add C to F using the operation $\oplus$. Then we obtain a set of smaller subtrees, see the figure. We perform such an operation in parallel for each hanging subtree containing a back-edge to v.

Stage 2. Elimination of cross-edges
In this stage the crucial operation is a parallel construction of a maximal matching between hanging subtrees. Construct an auxiliary graph whose nodes are

hanging subtrees; two subtrees are connected iff there is a crossing edge between them. Let H be a *maximal matching* in this graph. Consider each pair of subtrees T_i, T_j (in H) in parallel. There is a cross-edge (x, y) between these subtrees. Choose y to be the lowest node (in its subtree) with such a property. Then we take the cycle C consisting of this edge and paths from x and y to v. We add C to F applying the operation $\oplus$. The subtrees disconnect into a collection of smaller ones as in Stage 1.

The importance of maximality of the auxiliary matching H is rather subtle: it guarantees that subtrees having a cross-edge to other subtrees are disconnected from them, even if they do not participate in H.

The whole algorithm consists in finding in each bad component a good cut v and iterating Stage 1 followed by Stage 2. From Lemma 9.4.5 the number of iterations is logarithmic and we obtain the main result of this section.

Theorem 9.4.6
There is an NC-algorithm for testing the existence of and constructing a perfect matching (if there is any) in claw-free graphs.

It was shown by Dahlhaus and Karpinski [DK92] that the perfect matching problem is NC-reducible to a perfect matching problem for cubic graphs (regular graphs of degree 3). Using the algorithm above we can compute a spanning odd induced forest, and then we can present the graph as in Figure 9.11: dotted edges are non-forest edges. For bipartite cubic graphs Sharan and Wigderson [SW96] showed how to transform such a forest easily into a perfect matching. This was possible in this case since the graph formed by the dotted edges is a collection of even cycles and we can easily find a perfect matching for a family of disjoint even cycles. We refer to the paper of Sharan and Wigderson for details. For general graphs it is unclear how to do it.

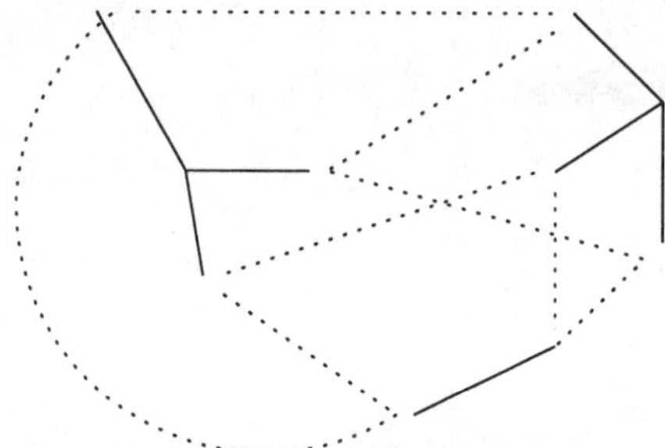

Fig. 9.11 The structure of a cubic graph: a pseudo-perfect matching consists of solid lines, and is an induced forest.

9.5 Bibliographic notes

The tree contraction method (used for matchings in trees) can be found in detail in Gibbons, Rytter [GR88]. The material on dense graphs is from Dahlhaus, Karpinski [DK87] and regular bipartite graphs were considered in Lev, Pippenger, Valiant [LPV81]. The processors' bound for regular bipartite graphs was recently changed, for degree 3, from $O(n)$ to $O(n\alpha(n)/\log n)$, see Sharan and Wigderson [SW96].

10

Families of intervals

A matching in a bipartite graph corresponds to a *system of distinct representatives* for a certain set family $\mathcal{I}$. In this chapter we consider the case when the family $\mathcal{I}$ is particularly natural, it is a family of intervals on the same line, and the intervals are indexed by elements of a set X. Formally there are integers $first_v$, $last_v$ such that

$$\mathcal{I} = \{\gamma_v : \gamma_v = [first_v \dots last_v], v \in X\}.$$

The *matching problem for convex graphs* is equivalent to the problem of finding a maximum cardinality system of distinct representatives for a family $\mathcal{I}$ of intervals.

A bipartite graph $G = (X, Y, E)$ is said to be *convex* iff Y is a subset of consecutive integers and

$$E = \{(v, k) : v \in X \text{ and } k \in [first_v \dots last_v]\}.$$

The set X is identified later with a set family $\mathcal{I}$.

In other word each vertex v can *see* ("through its edges") an interval of consecutive vertices in Y. An example of a convex bipartite graph is shown in Figure 10.1.

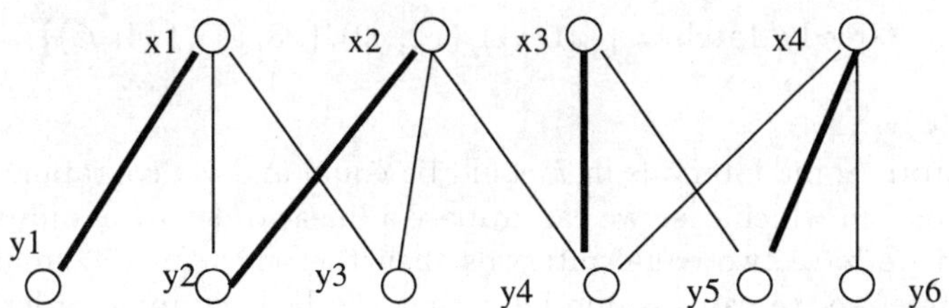

Fig. 10.1 A convex bipartite graph. For example $first_{x2} = 2$ and $last_{x2} = 4$. The corresponding set family is $\mathcal{I} = \{[1..3], [2..4], [4..5], [4..6]\}$. The *greedy matching* corresponds to the edges in bold.

Let $\mathcal{I}$ be a set of intervals. In the case of a convex graph corresponding to a set family $\mathcal{I}$ a matching is a set $\mathcal{M}$ of pairs (γ, x), where $\gamma \in \mathcal{I}$ and $x \in \gamma$. $\mathcal{M}$ should satisfy

$$\text{if } (\gamma, x), (\gamma', x') \in \mathcal{M} \text{ and } \gamma \neq \gamma' \text{ then } x \neq x'.$$

The maximum cardinality problem for convex bipartite graphs has some applications in scheduling. Consider the problem of scheduling a set X of unit time jobs to maximize the number of executed jobs. The job v has the *release time* $first_v$ and a *due time* $last_v$. In a given time unit only one job can be executed. The schedule maximizing the number of executed jobs corresponds directly to a maximum cardinality system of distinct representatives, or (in graph-theoretic terms) maximum cardinality matching in a corresponding graph.

Assume that the representation of the graph is given by its vertices and pairs $(first_v, last_v)$ for each vertex v. Hence the size of the representation is linear with respect to the number n of vertices (though the actual number of edges can be quadratic). Dekel and Sahni have designed an efficient parallel algorithm ($O(\log^2(n))$ time with $O(n)$ processors) by reducing the problem to some computations on trees, see [QD84].

10.1 The greedy sequential algorithm

The algorithm presented in this chapter is similar to that of Dekel and Sahni and is a parallel implementation of the *greedy* sequential algorithm. The matching produced by the greedy algorithm is denoted here by $\mathcal{M} = \text{GreedyMatch}(\mathcal{I})$. We also call it the *greedy matching.* The greedy matching $\mathcal{M}$ satisfies the following.

$\mathcal{M}$ is the lexicographically first matching among all matchings of maximum cardinality for $\mathcal{I}$. This means that the lexicographically first interval is matched to the first possible point, and after removing this interval and this point (from other intervals) the remaining matching has the same property.

Example Consider the family $\mathcal{I}$ of intervals from Figure 10.1. Then

$$\text{GreedyMatch} = \{(x1, y1), (x2, y2), (x3, y4), (x4, y5)\}.$$

Observation Some intervals in $\mathcal{I}$ could be equal and lexicographic order could be ambiguous, in which case we can impose a linear order by assuming additionally that if we have two equal intervals then the one with the smaller index is *earlier.* Therefore we can assume later that the lexicographic order for a given interval family $\mathcal{I}$ is unique.

Let $\text{FirstInterval}(\mathcal{I})$ denote the *lexicographically first* interval in $\mathcal{I}$. The greedy algorithm is presented below. Its correctness follows directly from the following obvious lemma.

Lemma 10.1.1
Let x be the first position of the interval $\gamma = \text{FirstInterval}(\mathcal{I})$. Then there is a maximum cardinality matching for $\mathcal{I}$ containing the pair (γ, x).

```
Algorithm Compute_Greedy_Match;
{Sequential greedy algorithm}
 M := ∅;
 while I ≠ ∅ do
    begin
    γ:= FirstInterval(I); x := first position of γ;
    M := M ∪ {(γ, x)};
    remove element x from all intervals in I;
    remove all empty intervals from I;
    end
 return GreedyMatch(I) = M;
```

10.2 Left-justification of set families

The basic idea in parallel implementation of the greedy sequential algorithm is to reduce the general case to a very special and easy case of *left-justified* set families $\mathcal{I}$. We introduce the crucial operation of *left-justification*. Denote

$$\begin{aligned} \mathrm{FirstPos}(\mathcal{I}) &= \min\{i : [i,j] \in \mathcal{I}\}; \\ \mathrm{LeftJustify}(\mathcal{I}) &= \{[\mathrm{FirstPos}(\mathcal{I}), j] : [i,j] \in \mathcal{I}\}. \end{aligned}$$

Hence FirstPos($\mathcal{I}$) is the first position of FirstInterval($\mathcal{I}$). Intuitively speaking the operation LeftJustify($\mathcal{I}$) extends all intervals in $\mathcal{I}$ to the left to the minimal position in all intervals. We are adding "artificial" elements to the interval. Hence the system of distinct representatives of LeftJustify($\mathcal{I}$) is usually not a system for $\mathcal{I}$, since some elements do not belong to the corresponding original intervals.

Example If $\mathcal{I} = \{[4..4], [5..8], [2..5], [15..20]\}$ then

$$\mathrm{LeftJustify}(\mathcal{I}) = \{[2..4], [2..8], [2..5], [2..20]\}.$$

A set $\mathcal{I}$ of intervals is said to be left-justified iff LeftJustify($\mathcal{I}$) $= \mathcal{I}$. Denote by LeftJustifiedMatching($\mathcal{I}$) the operation which returns the greedy matching for a left-justified family $\mathcal{I}$. The computation of maximum matchings for left-justified sets is the main operation in the parallel algorithm for the general case.

Lemma 10.2.1 (left-justification lemma)
Assume that intervals of $\mathcal{I}$ are lexicographically sorted and $\mathcal{I}$ is left-justified. Then LeftJustifiedMatching*($\mathcal{I}$) can be computed in $O(\log n)$ time with $O(n)$ processors.*

Proof Assume w.l.o.g. that the i-th interval is $1..l_i$. Denote
$d_i = \max(0, i - l_i)$ and $defect_i = \max\{\ d_k \ :\ k \leq i\}$.

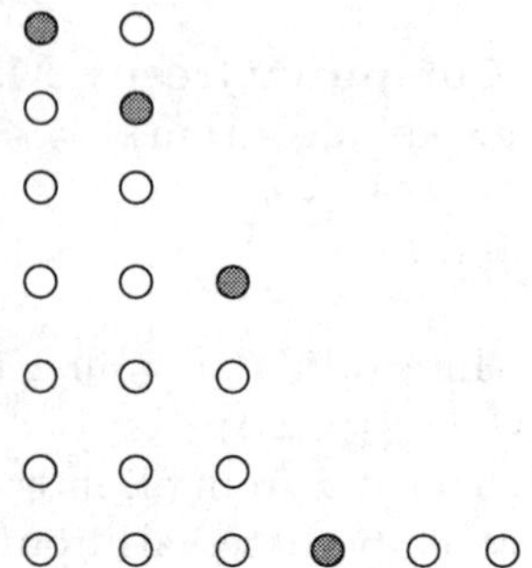

Fig. 10.2 A lexicographically sorted left-justified family of intervals. The points selected by the greedy matching are indicated in each interval.

Then the following claim follows from the defect theorem for bipartite graphs in Chapter 2. (Observe that, owing to the *greedy* approach, the matching related to the first i intervals does not depend on other intervals.)

Claim The number of intervals matched in GreedyMatch($\mathcal{I}$) with indices not exceeding i equals $i - defect_i$.

We can easily compute all values $defect_i$ using a parallel prefix computation. Then the i-th interval is matched to the point $(i-1-defect_{i-1})+1$, if this point is inside the i-th interval. If not then the i-th interval is not matched.

For example, the seventh interval in Figure 10.2 receives the point 4. All that can be done in parallel for each interval.

□

The main "trick" in the algorithm for the general case consists in inserting additional "artificial" points to make intervals left-justified. Then the operation LeftJustifiedMatching is applied. However, some intervals can receive "artificial" elements which originally were not contained in these intervals and the actual matching $\mathcal{M}'$ would be invalid with respect to the original intervals, so a kind of *repairing* $\mathcal{M}'$ (described below) should be applied. The operation REPAIR($\mathcal{M}$)′ works under the following assumptions:

1. We have two families of intervals $\mathcal{I}$ and $\mathcal{I}'$, where $\mathcal{I}'$ is left-justified and FirstPos($\mathcal{I}'$) $\leq$ FirstPos($\mathcal{I}$).
2. We are given a greedy matching $\mathcal{M}$ of $\mathcal{I}$.
3. We are given a greedy matching $\mathcal{M}'$ of LeftJustified($\mathcal{I} \cup \mathcal{I}'$).

The value of REPAIR is a greedy matching of $\mathcal{I}' \cup \mathcal{I}$.

Lemma 10.2.2 (key lemma)
The operation REPAIR($\mathcal{M}'$) *can be computed in* $O(\log n)$ *time with* $O(n)$ *processors if the matching* $\mathcal{M}$ *is given, where* $n = |\mathcal{I}'| + |\mathcal{I}|$.

Proof We can remove from $\mathcal{I}$ intervals which are not matched in $\mathcal{M}$, so we can assume later that each interval in $\mathcal{I}$ is matched w.r.t. $\mathcal{M}$.

If $\mathcal{M}'$ is valid for $\mathcal{I} \cup \mathcal{I}'$ then $\mathcal{M}'$ is the required final result.

Otherwise consider any invalid pair $(\gamma_1, i_1) \in \mathcal{M}'$: it is invalid if $i_1 \notin \gamma_1$, and the point i_1 is an *artificial* point added to γ_1 in the process of *left-justification*. Let i_2 be the point to which γ_1 was matched in $\mathcal{M}$. If this point is *free* in $\mathcal{M}'$ then we can replace i_1 by i_2 and have a valid matching position for γ_1.

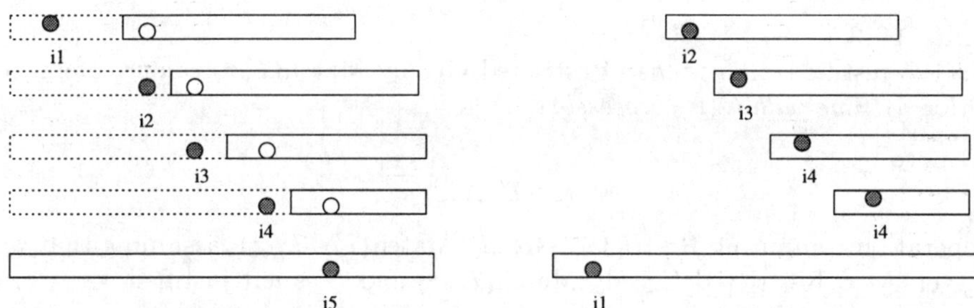

Fig. 10.3 An illustration of the operation of repairing the matching $\mathcal{M}'$ locally on a given *chain*, in this case $k = 4$, and the second case holds. The circles in bold in the left figure correspond to the given matching $\mathcal{M}'$ of LeftJustified$(\mathcal{I}' \cup \mathcal{I})$. The circles in bold on the right correspond to the resulting matching REPAIR$(\mathcal{M}')$.

Otherwise i_2 is matched to some interval γ_2. In this way we have a sequence of intervals $\gamma_1, \gamma_2, \ldots, \gamma_k$ belonging to $\mathcal{I}$ and the sequence of points $i_1, i_2, \ldots, i_{k+1}$, see Figure 10.3, such that for each $1 \le t \le k$:

1. γ_t is matched to i_t in $\mathcal{M}'$.
2. γ_t is matched to i_{t+1} in $\mathcal{M}$.

We call such a maximal (w.r.t. inclusion) sequence of intervals a *chain*. Such chains are being *repaired* locally. There are two cases to consider for a given chain:

Case 1: i_{k+1} is *free* in $\mathcal{M}'$.
We *shift* the matching $\mathcal{M}'$ locally: γ_t gets the (valid) point i_{t+1} for each $1 \le t \le k$.

Case 2: i_{k+1} is matched to an interval $\gamma \in \mathcal{I}'$.
In this case we give to γ the point i_1, which is valid for γ since $\mathcal{I}'$ is left-justified. Then we proceed in the same way as in Case 1.

We process each chain (independently) in parallel in the way described above. Some simple parallel processing of lists is required. Altogether logarithmic time with a linear number of processors is enough.

□

10.3 The parallel algorithm for the general case

Let $\mathcal{I}$ be a family of intervals and τ be an interval (usually $\tau \notin \mathcal{I}$).

Denote by Restricted_Greedy_Match($\mathcal{I}, \tau$) the greedy matching for $\mathcal{I}$ with the restriction that all representatives belong to τ. In other words it is a greedy matching for the *restricted* family $\mathcal{F} = \{\gamma \cap \tau \;:\; \tau \in \mathcal{I}\}$, with the original names of intervals preserved.

We say that $\mathcal{I}$ is left-justified w.r.t. τ iff FirstPos($\mathcal{I}$) $\leq$ FirstPos(τ).

The following lemma is an obvious extension of Lemma 10.2.1.

Lemma 10.3.1
If $\mathcal{I}$ *is* left-justified *w.r.t.* τ *then* Restricted_Greedy_Match($\mathcal{I}, \tau$) *can be computed in* $O(\log n)$ *time with* $O(n)$ *processors.*

Denote by

$$\text{Merge}(\mathcal{I}', \mathcal{M}, \tau)$$

the operation computing Restricted_Greedy_Match($\mathcal{I}' \cup \mathcal{I}$, τ), assuming that we are given $\mathcal{M}$ = Restricted_Greedy_Match($\mathcal{I}', \tau$) and $\mathcal{I}'$ is left-justified w.r.t. τ.

Lemma 10.3.2 (extending match lemma)
Assume we have two families of intervals $\mathcal{I}$ *and* $\mathcal{I}'$, *where* $\mathcal{I}'$ *is left-justified w.r.t. a given interval* τ. *Assume that we are given* $\mathcal{M}$ = Restricted_Greedy_Match($\mathcal{I}, \tau$). *Then* Merge($\mathcal{I}', \mathcal{M}, \tau$) *can be computed in* $O(\log n)$ *time with* $O(n)$ *processors, where* $n = |\mathcal{I}'| + |\mathcal{I}|$.

Proof An easy consequence of Lemma 10.2.2 and Lemma 10.3.1.

□

Let $\pi = (f_1, f_2, \ldots, f_k)$ be the sorted sequence (in increasing order) of the first positions of intervals in $\mathcal{I}$. Assume w.l.o.g. that k is a power of two.

We construct a tree T_{int} of intervals; its leaves are intervals

$$[f_1..f_2 - 1], [f_2..f_3 - 1], \ldots, [f_{n-1}..f_n - 1], [f_n..\infty].$$

The parent of two adjacent intervals results by merging them together, see Figure 10.4. Denote by LeftPart(τ), RightPart(τ) the son-intervals of a given interval τ corresponding to an internal node of the interval tree.

The set family $\mathcal{I}$ is processed recursively. At the bottom of the recursion are subfamilies consisting of all intervals which start at the same position f_i. The main point is that these subfamilies are left-justified.

We can view subfamilies of intervals as organized in a tree. At a given level of recursion the subfamilies corresponding to this level are processed at the same time. The tree structure for an example set family is illustrated in Figure 10.5.

Example Consider the following family $\mathcal{I}$ of intervals, sorted lexicographically:

$$I_1 = [1..2],$$

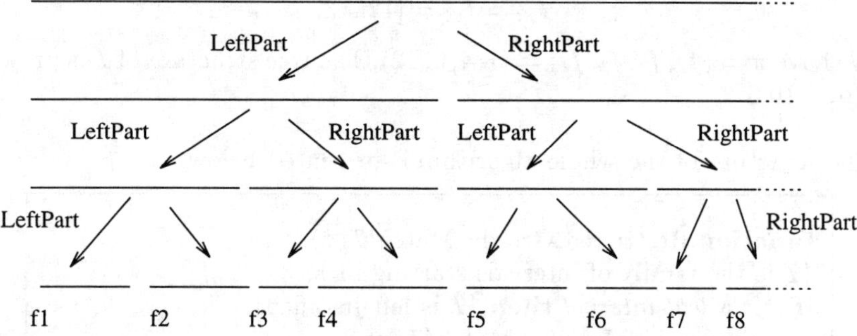

Fig. 10.4 The tree T_{int} of intervals, the dots meaning "going to infinity". The bottom intervals are *leaf-intervals*; each *internal* interval τ has two *sons*: LeftPart(τ), RightPart(τ).

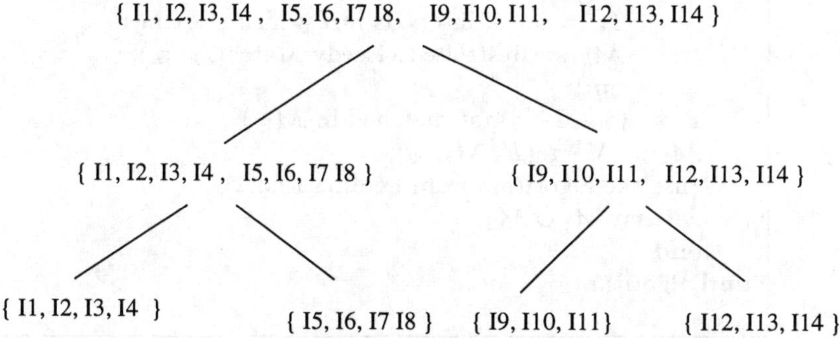

Fig. 10.5 The tree structure of the family of intervals from the example.

$$I_2 = [1..3],$$

$$I_3 = I_4 = [1..4],$$

$$I_5 = [4..4],$$

$$I_6 = [4..8],$$

$$I_7 = [4..10],$$

$$I_8 = [4..11],$$

$$I_9 = [9..10],$$

$$I_{10} = [9..11],$$

$$I_{11} = [9..13],$$

$$I_{12} = [12..12],$$

$$I_{13} = I_{14} = [12..13].$$

We have $\pi = (f_1, f_2, f_3, f_4) = (1, 4, 9, 12)$. The tree structure of $\mathcal{I}$ is presented in Figure 10.5.

The structure of the whole algorithm is presented below.

```
function Restricted_Greedy_Match(I, τ);
{I is the family of intervals starting in τ, τ ∈ T_int }
if τ is a leaf-interval then {I is left-justified}
    return LeftJustifiedMatch(I, τ)
{use the algorithm from Lemma 10.2.1}
    else begin
    τ1 := LeftPart(τ); τ2 := RightPart(τ);
    for k = 1,2 do in parallel
            begin
            I_k := set of intervals in I which start in τ_k
            M_k := Restricted_Greedy_Match(I_k, τ_k);
            end;
    I' := {γ ∈ I_1: γ not matched in M_1 } ;
    M_2 := Merge(I', M_2, τ_2);
    {use the algorithm from Lemma 10.3.2}
    return M_1 ∪ M_2;
    end
end algorithm
```

The greedy matching for the initial family $\mathcal{I}$ is computed by calling the algorithm Restricted_Greedy_Match$(\mathcal{I}, \tau)$ for the *root interval* $\tau = [f_1..\infty]$. The depth of the recursion is logarithmic. One level of recursion is completed in logarithmic time with $O(n)$ processors.

In this way we have shown constructively the following theorem.

Theorem 10.3.3
The maximum cardinality matching problem for convex bipartite graphs can be solved in $\log^2 n$ *time with* $O(n)$ *processors on a* CREW PRAM.

10.4 Bibliographic notes

The material in this chapter is mostly based on Quinn, Deo [QD84].

11

Parallel algorithms for f-matchings

In this chapter we present randomized and deterministic NC-algorithms for maximum and (inclusion) maximal *f-matchings* (which are natural generalizations of matchings). In particular each (usual) matching is a 1-matching. Let G be an undirected graph with an integer capacity $f(v)$ associated with each vertex v. A subset M of the set of edges of G is an *f-matching* if and only if each vertex v in G is incident to at most $f(v)$ edges in M. An f-matching is *maximal* if and only if it is not properly contained in another f-matching.

An example of a maximal f-matching containing 10 edges is illustrated in Figure 11.1; the maximum cardinality f-matching in this case has 11 edges. The parallel computation of a maximal f-matching is generally more difficult than that of a maximal matching. Obviously, a maximal f-matching can be computed in linear time by a greedy sequential algorithm. Unfortunately we do not know how to parallelize this algorithm. In Chapter 7 we presented an RNC-algorithm for maximal matchings. By generalizing this randomized algorithm we can present a simple randomized algorithm for maximal f-matching which runs in $\mathcal{O}(\log^3 n)$ time with $\mathcal{O}(n+m)$ processors.

The deterministic NC-algorithm is much more complicated. It is based on the Israeli and Shiloach deterministic algorithm for maximal matchings, see [IS86]. It was generalized in [DGL93] to maximal f-matching.

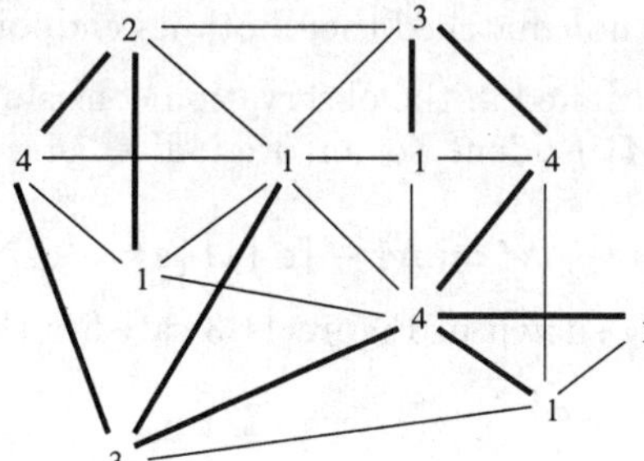

Fig. 11.1 An example graph with capacities placed in the nodes and a possible (inclusion) maximal f-matching $\mathcal{M}$. The edges of $\mathcal{M}$ are in bold. $\mathcal{M}$ is not a maximum cardinality f-matching.

11.1 *RNC*-algorithm for maximum f-matchings

We reduce the problem of maximum cardinality f-matching to the maximum cardinality of the (usual) 1-matching by enlarging the graph $G = (V, E)$.

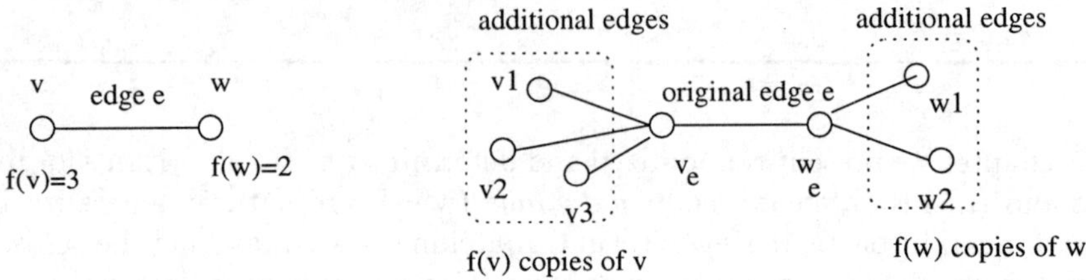

Fig. 11.2 Local construction of the part of the graph G' related to the edge $e = (v, w)$.

The new graph is $G' = (V', E')$, where:

- V' consists of $f(v)$ copies $v_1, v_2, \ldots, v_{f(v)}$ of each vertex $v \in V$ and *special vertices* v_e for each vertex v and each edge e incident to v.
- E' consists of edges of the form (v_e, v_w) corresponding to edges $e = (v, w)$ and called *original edges.* Additionally for each copy v_i of the vertex v we have an edge connecting v_i to each special vertex v_e, where e is any edge incident to v.

A part of the graph G' relevant to a vertex v is illustrated in Figure 11.2 and Figure 11.3 shows globally a transformation of an example graph G.

Observation Each maximum cardinality matching of G' can be easily transformed in parallel to a maximum cardinality matching $\mathcal{M}$ satisfying the following condition:

(*) If an original edge is not matched then both its endpoints are covered by $\mathcal{M}$.

The transformation mentioned in the observation consists in exchanging (in parallel) any edge $e' \in \mathcal{M}$ incident to an original edge e violating (*), that is performing

$$\mathcal{M} := \mathcal{M} - \{e'\} \cup \{e\}.$$

Observe that a matching shown in Figure 11.3 satisfies the condition (*).

Theorem 11.1.1
Let $P(n)$ be the number of processors sufficient for computing a maximum matching in $O(\log^k n)$ average time for graphs having n vertices. Then a maximum f-matching can computed in $O(\log^k n)$ time with $O(P(n \cdot \max(f))$, where $\max(f) = \max\{f(v) \;:\; v \in V\}$.

Proof The problem is reduced to the usual matching by transforming G into G'. We construct the graph G' and compute a maximum matching $\mathcal{M}$ in G' satisfying the condition (*). The thesis is reduced to the following claim.

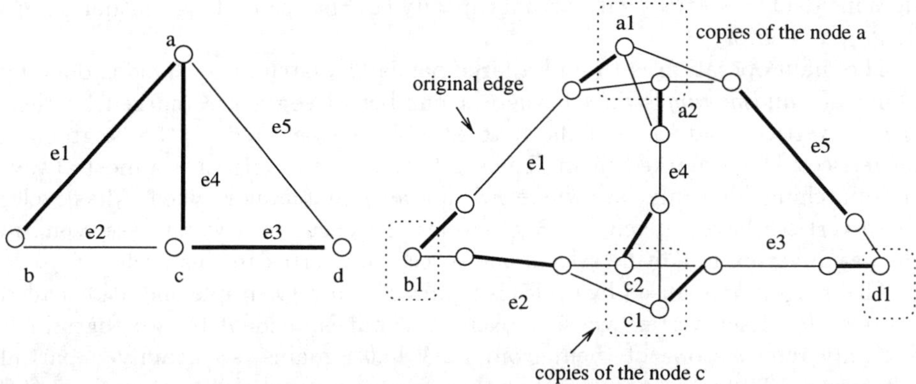

Fig. 11.3 Transformation of an example graph G (on the left) to G' (on the right). Assume $f(a) = 1$, $f(b) = 2$, $f(c) = 2$, $f(d) = 1$. A maximum matching $\mathcal{M}'$ in G' is indicated in bold; it corresponds to maximum f-matching $\mathcal{M}$ in G consisting of all original edges in G' with both endpoints free.

Let $\mathcal{M}_f$ be the set of original edges in G' which are not in $\mathcal{M}$. Then $\mathcal{M}_f$ is a maximum cardinality f-matching for G.

It is easily seen that $\mathcal{M}_f$ is an f-matching. We prove its maximality by observing the following obvious equality (each original edge e of G' is in $\mathcal{M}$ or it is in $\mathcal{M}_f$ and then two nonoriginal edges incident in G' to e are in $\mathcal{M}$):

$$|\mathcal{M}| = |\mathcal{M}_f| + |E|.$$

Hence

$$|\mathcal{M}_f| \le |\mathcal{M}| - |E|.$$

On the other hand, for any f-matching $\mathcal{M}'$ we can construct a matching $\mathcal{M}$ in G' such that $|\mathcal{M}_f| = |\mathcal{M}'|$. So f-matching $\mathcal{M}_f$ corresponding to a maximum matching $\mathcal{M}$ of G' is a maximum f-matching of G.

□

Observation Surprisingly if $\mathcal{M}$ is an inclusion maximal matching of G' then $\mathcal{M}_f$ is not necessarily a maximal f-matching in G. For example, $\mathcal{M}$ can consist of all original edges, in this case $\mathcal{M}_f = \emptyset$'.

11.2 Simple RNC-algorithm for maximal f-matchings

We present first a simple randomized algorithm, which is much simpler than the only known deterministic NC-algorithm presented in the next section.

We extend here the method of *handshaking* developed in Chapter 4 and used in Chapter 7. The main difference is that we allow a person to have more than two hands.

The whole structure of the randomized algorithm is very simple: it finds a partial f-matching F, deletes the edges of F, updates capacities and then removes edges incident to vertices with actual capacity equal to zero. This continues until the graph is empty.

The main operation is to find a large partial f-matching F. This is done by a kind of *handshaking strategy* where a number of edges are marked by their incident vertices and some of the marked edges are selected by the neigbors of the vertices which marked them. The set F of selected vertices is almost always an f-matching. The only case where F is not an f-matching is when F has cycles whose vertices have capacity 1. Such cycles can occur in a worst case scenario when each vertex in a cycle selects one succeeding vertex in this cycle.

The irregularity caused by cycles is repaired by a very simple operation called *repair cycles.* Each vertex whose capacity is 1 but is incident to two edges in F randomly removes one of them from F. What remains is certainly a partial matching F. The main point is that there is a large probability that the set F is large afterwards.

In the algorithm and subsequently, $\deg(H, v)$ means the number of edges in H incident to v. In particular if H is the set of all edges of the graph we simply write $\deg(v)$.

```
Algorithm MaxMatch(G, f);
    M := ∅;
    repeat
        handshaking1:
            for each vertex v do in parallel
                mark at random ⌈½f(v)⌉ incident edges;
        handshaking2:
            for each vertex v do in parallel
                Select at random max{⌊½f(v)⌋, 1} edges
                marked by its neighbors;
            F := the set of selected edges;
        repair cycles:
            for each vertex v do in parallel
                if f(v) = 1 and (v has two incident edges in F)
                    then randomly delete one of them;
            {F is now an f-matching}
            M := M ∪ F;
            E := E \ F;
        clean-up:
            for each vertex v do in parallel
                if deg(F, v) = f(v)
                    then remove all edges incident to v;
                f(v) := min{deg(E, v), f(v) − deg(F, v)}
    until G has no edges;
    return M;
end.
```

The analysis of the algorithm MaxMatch is based on the *good-edges* lemma from Chapter 2.

Associate with each vertex v of G a number $\varphi(v)$. We say that a neighbor w of v is a *nice neighbour* if and only if $\varphi(v) \leq \varphi(w)$. The vertex v is *good* if and only if at least $\frac{1}{3}$ of its neighbors are nice. The edge is called *good* if and only if at least one of its endpoints is good (see Fig. 1). Vertices which are not good are called *bad* vertices and edges which are not good are called *bad* edges. In Chapter 2 we proved the following lemma

Lemma 11.2.1
At least half of the edges of G are good.

The crucial point in our algorithm is to reduce capacities for a big proportion of the so-called good edges by a constant fraction in each iteration. This guarantees the logarithmic number of iterations since at least half of the edges of the graph are always good.

We take here

$$\varphi(v) = \frac{f(v)}{\deg(v)}.$$

The lemma below says that good vertices receive many edges on average from their neighbors. This can also be seen at the top graph in Fig. 11.2.

Lemma 11.2.2
Consider a single performance of the main iteration in MaxMatch *applied to a graph G. The probability that a good vertex v of positive degree in G and positive capacity $f(v)$ has degree $\geq \frac{1}{7}f(v)$ in F directly after the handshaking1 and handshaking2 stages in the iteration is bounded by a positive constant from below.*

Proof Let v be a good vertex, and let $d = \deg(v)$. There are k neighbours $v_1, \ldots, v_k$ of v such that $k \geq \lfloor \frac{1}{3}d \rfloor + 1$, and $(1/d)f(v) \leq f(v_i)/\deg(v_i)$ for $i = 1, \ldots, k$. The probability that v_i marks the edge (v_i, v) in the handshaking1 stage is at least $f(v)/2d$. It follows that the expected number of edges incident to v marked by all neighbors of v in this stage is at least $\frac{1}{6}f(v)$.

In [HR89] the following variant of Chernoff's bound has been derived: let $X_1, X_2, \ddots, X_k$ be independent 0–1 variables, and let M be the expected value of $S = X_1 + X_2 + \ddots + X_k$. Then

$$Pr(S \leq (1-\epsilon)M) \leq \left(\frac{\exp(\epsilon)}{(1+\epsilon)^{1+\epsilon}}\right)^M, \quad 0 \leq \epsilon \leq 1.$$

In our case, we have $M \geq \frac{1}{6}f(v)$. Substitute $(\frac{1}{7})$ for ϵ and observe that

$$\left(\frac{\exp(\epsilon)}{(1+\epsilon)^{1+\epsilon}}\right)^{\frac{1}{6}} < 1.$$

It follows by Chernoff's bound that the probability that the number of edges incident to v marked by all neighbors in the first stage is less than $\frac{1}{7}f(v)$ is bounded from above by a constant smaller than one. This completes the proof since in the repair-cycle stage of the iteration, vertex v chooses $\max\{\lfloor\frac{1}{2}f(v)\rfloor, 1\}$ edges among the above marked edges.

□

Lemma 11.2.3
The expected number of main iterations in the algorithm MaxMatch is $\mathcal{O}(\log^2 n)$.

Proof Let e denote the number of edges of the current graph G in MaxMatch. To prove the lemma it is enough to show the following claim.

Claim There exists a constant d such that after $d\lceil\log n\rceil$ iterations, the expected number of edges in G is not greater than $\frac{1}{2}e$.

The proof of this fact, where d is specified later, is as follows.

Suppose that after $d\lceil\log n\rceil$ iterations the expected number of edges in G is at least $\frac{1}{2}e$. Consider any of the above iterations. An edge in the current G in the iteration is said to be *very good* if it is good and the capacity of at least one of its good endpoints is decreased at least by one-seventh. Combining Lemma 11.2.1 with Lemma 11.2.2, we conclude that directly after the repair-cycle phase in the iteration, the expected number of edges that can become very good if they survive the clean-up phase is at least a constant fraction of $\frac{1}{2}e$, say $\geq b \cdot e$ for some positive constant b.

The clean-up phase decreases the above number by at most one-half to at least $\frac{1}{2}be$. Therefore we conclude that during the $d\lceil\log n\rceil$ iterations, the expected number of occurrences of good edges is at least $\frac{1}{2}bed\lceil\log n\rceil$. On the other hand, by Lemma 11.2.2, there exists a constant c such that a vertex v can occur as a good vertex, which has a positive capacity and loses at least one-seventh of its capacity in at most $c\lceil\log n\rceil$ iterations.

It follows that an edge can occur in at most $2c\lceil\log n\rceil - 1$ iterations as a very good edge, being deleted from G in the last iteration. Set d to be $= 4\frac{c}{b}$. Then, the expected total number of occurrences of very good edges in the $d\lceil\log n\rceil$ iterations is greater than $e(2c\lceil\log n\rceil - 1)$. We conclude that the expected number of edges in G after the $d\lceil\log n\rceil$ iterations is 0, which contradicts our assumption.

□

The lemma directly implies the following result.

Theorem 11.2.4
We can compute a maximal f-matching in $\mathcal{O}(\log^3 n)$ expected time with $\mathcal{O}(n+m)$ processors.

11.3 Deterministic NC-algorithm for maximal f-matchings

We present a deterministic parallel algorithm for maximal f-matching which is a generalization of the algorithm due to Israeli and Shiloach for maximal matching,

[IS86]]. The analysis of our algorithm is different from that of the Israeli–Shiloach algorithm in crucial points (e.g. in the proof of the key lemma).

The algorithm due to Israeli and Shiloach constructs a maximal matching in a graph on n nodes and m edges in time $\mathcal{O}(\log^3 n)$ time on an CRCW PRAM with $\mathcal{O}(n+m)$ processors, [IS86]. In this subsection we present an advanced generalization of the above algorithm to include maximal f-matching achieving the same asymptotic resource bounds. The generalized algorithm consists of several procedures. In order to present them we need the following

Let $\mathcal{N}$ denote the set of natural numbers. For each vertex $v \in V$ we define its weight $w(G,v,f)$ with respect to f as follows:

$$w(G,v,f) = \min_{s\in\mathcal{N}}\{s \mid 2^s f(v) \geq d(G,v)\}.$$

Let $W(G,f) = \max_{v\in V} w(G,v,f)$. The number $W(G,f)$ will be called the weight of the graph G with respect to f.

Observe that if $W(G,f)=0$ then all edges of the graph belong to a maximal f-matching.

In the procedure REDUCE(G,f,G') we are given a simple graph $G=(V,E)$, and a matching function $f : V \to \mathcal{N}$ such that $W(G,f) \geq 1$. We assume that $d(G,v), w(G,v,f)$ for all $v \in V$ and $W(G,f)$ are computed.

procedure REDUCE(G,f,G')
construct an auxiliary graph H induced by all those edges of G for which at least one endpoint is an active vertex (called later real edges);
compute connected components of H;
label edges of each connected component 1 if its all active vertices are safe; call such components safe;
let H' be a subgraph of H containing only nonsafe components;
for every vertex of odd degree in H' add an edge to an introduced vertex u;
find an Eulerian circuit in each connected component C of H';
label the edges of each component C of H' with 0 and 1 in the following way:
(1) if C contains the introduced vertex u **then** starting at u
trace the Eulerian cycle and label the edges 0 and 1 alternately;
if the number of real edges labeled 0 is larger than the number of real edges labeled 1 then exchange zeros for ones and ones for zeros;
(2) if C does not contain u **then** find in C a vertex x
which is a neighbor of a nonsafe vertex y;
starting from the edge (x,y) trace the cycle and label the edges 1 and 0 alternately ((x,y) is labeled 1);
if the length of the cycle is odd then exchange the label of (x,y) for 0;
$G' \leftarrow$ the graph obtained by the removal of all real edges labeled with 0 from G;
end

The procedure returns a subgraph $G' = (V, E')$ of the graph G such that $W(G', f) \leq W(G, f) - 1$.

A vertex v is said to be an active one in G with respect to a matching function f if and only if $2^{W(G,f)-1}f(v) \leq d(G, v) \leq 2^{W(G,f)}f(v)$. An active vertex v is called *safe* if $d(G, v) = 2^{W(G,f)-1}f(v)$. A procedure REDUCE, defined above, reduces G (removing some of its edges) to a graph G' such that $W(G', f) \leq W(G, f) - 1$.

Lemma 11.3.1
If $W(G, f) \geq 1$ then $W(G', f) \leq W(G, f) - 1$.

Proof First of all let us observe that if $w(G, v, f) \leq W(G, f) - 1$, for some $v \in V$, then also $w(G', v, f) \leq W(G, f) - 1$.

Consider now a vertex v for which $w(G, v, f) = W(G, f)$. Then

$$2^{W(G,f)-1}f(v) < d(G, v) \leq 2^{W(G,f)}f(v)$$

and v is a nonsafe vertex. Hence $d(G', v) \leq \lceil \frac{1}{2}d(G, v) \rceil \leq 2^{W(G,f)-1}f(v)$. This implies $W(G', f) \leq W(G, f) - 1$.

□

Lemma 11.3.2
Let $W(G, f) \geq 2$. If v is an active vertex in G then v remains active in G'.

Proof It suffices to show that if v is an active vertex in G then $2^{W(G,f)-2}f(v) \leq d(G', v) \leq 2^{W(G,f)-1}f(v)$. Let us consider three cases:

1. v belongs to a connected component of H in which all active vertices are safe. Then
$$d(G', v) = d(G, v) = 2^{W(G,f)-1}f(v).$$
2. v is in the same connected component of H' as the introduced vertex u. The procedure REDUCE reduces the input graph in such a way that
$$\lfloor \frac{1}{2}d(G, v) \rfloor \leq d(G', v) \leq \lceil \frac{1}{2}d(G, v) \rceil.$$
Hence
$$2^{W(G,f)-2}f(v) \leq d(G', v) \leq 2^{W(G,f)-1}f(v).$$
3. v is in a connected component of H' not containing u. If $v \neq y$ (see description of the procedure) then
$$d(G', v) = \frac{d(G, v)}{2}.$$
If $v = y$ then v is a nonsafe vertex. Hence
$$d(G, v) \geq 2^{W(G,f)-1}f(v) + 2.$$

If the Eulerian circuit in the connected component contained v has odd length then

$$d(G', v) = \frac{d(G, v)}{2} - 1 \geq 2^{W(G,f)-2} f(v),$$

otherwise

$$d(G', v) = \frac{d(G, v)}{2}.$$

Hence always

$$2^{W(G,f)-2} f(v) \leq d(G', v) \leq 2^{W(G,f)-1} f(v).$$

□

We define a procedure f-MATCHING which computes some "large" f-matching M for a given input graph G.

```
procedure f-MATCHING(G,f,M)
{given a graph G = (V, E) and a matching function f : V → N
procedure returns a large f-matching for G}
  i := 0; G_0 := G;
  repeat
      for all vertices v ∈ V do in parallel
            compute d(G_i, v); compute w(G_i, v, f);
      compute W(G_i, f); j := i;
      if W(G_i, f) ≥ 1 then
            REDUCE(G_i, f, G_{i+1}); i := i + 1;
  until j = i;
  M ← the set of edges of the graph G_i;
end procedure
```

Lemma 11.3.3
The set of edges M computed by the procedure f-MATCHING is an f-matching in the input graph G.

Proof Let $i_{\max}$ be the maximal value of the variable i. The weight of the graph $G_{i_{\max}}$ with respect to the function f is 0. Then for each vertex $v \in V, d(G_{i_{\max}}, v) \leq f(v)$. This implies that M is an f-matching in G.

□

Let M be an f-matching in a graph G. The procedure MODIFY, defined below, deletes the edges of M from the graph producing a graph K and next computes a matching function h such that any maximal h-matching of K extended with the edges of the set M is a maximal f-matching in G.

In the procedure MODIFY(G, f, M, K, h) we are given a graph $G = (V, E)$, its matching function f and an f-matching M in G.

The procedure returns a subgraph $K = (V, E')$ of G and a matching function h for K such that any maximal h-matching H extended with the edges of M is a maximal f-matching of the graph G.

```
procedure MODIFY(G, f, M, K, h);
    M' ← {(u, v) ∈ E | (u, v) ∈ M
                         or u is incident to f(u) edges in M
                         or v is incident to f(v) edges in M};
    E' := E \ M'; K := (V, E');
    for v ∈ V do in parallel
        if d(K, v) = 0 then h(v) := 1
          else h(v) := f(v) − |{(v, u) | (v, u) ∈ M}|;
        return K, h;
end
```

Lemma 11.3.4
Let H be a maximal h-matching in K. Then $H \cup M$ is a maximal f-matching in G.

Proof It is sufficient to observe that:

- E' consists of all possible edges which can extend the f-matching M, and
- for each nonisolated vertex v in H its new matching value $h(v)$ is equal to the old matching value $f(v)$ decreased by the number of edges belonging to the f-matching M.

□

Now we are ready to write the entire algorithm for finding maximal f-matchings in graphs:

```
Algorithm Maximal-f-Matching(G, f)
{given a graph G = (V, E), and a matching function f: V → N}
{algorithm computes a maximal f-matching MaxM in the graph G}
i := 0; G0 := G;
f0 := f; {G0 = (V0, E0) = G = (V, E)}
MaxM := ∅;
    while |Ei| > 0 do
        f-MATCHING(Gi, fi, Mi); MaxM := MaxM ∪ Mi;
        MODIFY(Gi, fi, Mi, Gi+1, fi+1); i := i + 1;
end
```

Consider a graph G and its matching function f. Let A be a subset of the set of vertices of the graph. By cost(G, A, f) we will denote a cost of the set A with respect to the function f defined as follows:

$$\mathrm{cost}(G, A, f) = \sum_{\substack{u \text{ is non-isolated} \\ \text{in } A}} f(u).$$

If A is empty or contains only isolated vertices, we assume $\mathrm{cost}(G, A, f) = 0$.

Lemma 11.3.5

Let G be a graph, f its matching function and C a vertex cover of the graph G. Let M be the f-matching which is the result of the call f-MATCHING(f, G, M). Finally, let K, h be the graph and its matching function, respectively, obtained as the result of the call MODIFY(G, f, M, K, h). Then there exists a vertex cover A in the graph K such that $\mathrm{cost}(K, A, h) \leq \frac{5}{6}\mathrm{cost}(G, C, f)$.

Proof Let us consider the application of the procedure f-MATCHING to the graph G. Let i_{max} be the maximum value of the variable i. If $i_{\mathrm{max}} = 0$ then all edges of the graph belong to M. In this case it suffices to take as the set A simply the empty set. Let us assume now that $i_{\mathrm{max}} > 0$. Let B denote a set of all active vertices in the graph $G_{i_{\mathrm{max}}-1}$. Each edge $e \in E$ has either both endpoints in $V \setminus B$ or at least one of its endpoints belong to B. If both endpoints belong to $V \setminus B$ then naturally e is an edge in the graph $G_{i_{\mathrm{max}}}$ and hence it is in M. This and the fact that each active vertex in G_j, for each $j < i_{\mathrm{max}-1}$, remains active in $G_{i_{\mathrm{max}-1}}$ (Lemma 11.3.4) imply that B is a vertex cover of the graph K.

Let us observe that

$$\mathrm{cost}(G, B, f) = \sum_{u \in B} f(u) \leq \sum_{u \in B} d(G_{i_{\mathrm{max}-1}}, u).$$

Let l denote the number of edges in $G_{i_{\mathrm{max}-1}}$ with exactly one endpoint in B and k the number of edges with both endpoints in B.

Then $\mathrm{cost}(G, B, f) \leq l + 2k$. The procedure REDUCE reduces the graph $G_{i_{\mathrm{max}-1}}$ in such a way that the f-matching M contains at least $\frac{1}{3}(l + k)$ edges incident with vertices in B.

Each connected component of H in which all active vertices are safe gives all its edges to M. The connected component in H' containing the introduced vertex u gives at least half of its real edges to M. If a connected component C of H' does not contain u and has e edges then it gives $\lfloor \frac{e}{2} \rfloor \geq \frac{e}{3}$ edges to M ($e > 1$, because C contains a safe active vertex). Since $\frac{1}{3}(l + k) \geq \frac{1}{6}(l + 2k)$, we have the following:

$$\begin{aligned}
\mathrm{cost}(K, B, h) &\leq \mathrm{cost}(G, B, f) - \tfrac{1}{3}(l + k) \\
&\leq \mathrm{cost}(G, B, f) - \tfrac{1}{6}(l + 2k) \\
&\leq \mathrm{cost}(G, B, f) - \tfrac{1}{6}\mathrm{cost}(G, B, f) \\
&\leq \tfrac{5}{6}\mathrm{cost}(G, B, f).
\end{aligned}$$

Let us now consider two cases:

Case 1 $\mathrm{cost}(G, B, f) \leq \mathrm{cost}(G, C, f)$. If we take as the set A the set B then

$\mathrm{cost}(K, A, h) \leq \frac{5}{6}\mathrm{cost}(G, C, f)$.

Case 2 $\mathrm{cost}(G, B, f) > \mathrm{cost}(G, C, f)$. Let us observe that M contains at least $\frac{1}{6}\mathrm{cost}(G, B, f)$ edges. If we take C as the set A the the following holds:

$$\begin{aligned} \mathrm{cost}(K, A, h) &\leq \mathrm{cost}(G, C, f) - \tfrac{1}{6}\mathrm{cost}(G, B, f) \\ &\leq \mathrm{cost}(G, C, f) - \tfrac{1}{6}\,\mathrm{cost}(G, C, f) \\ &\leq \tfrac{5}{6}\,\mathrm{cost}(G, C, f). \end{aligned}$$

□

Theorem 11.3.6

Let $G = (V, E)$ be an n-vertex graph with m edges and let f be its matching function. A maximal f-matching in G can be computed in $\mathcal{O}(\log^3 n)$ time using a CRCW PRAM *with $\mathcal{O}(n + m)$ processors.*

Proof It follows directly from Lemma 11.3.3 and Lemma 11.3.4 that if the algorithm MAXIMAL-f-MATCHING stops then MaxM is a maximal f-matching in G. The complexity of the procedure REDUCE depends on the complexity of computing connected components and Eulerian cycles. This can be done in time $\mathcal{O}(\log^2 n)$ with a linear number of processors, see [GR88]. It follows from Lemma 11.3.2 that the number of iterations of the repeat loop cannot be larger than $\lceil \log(n-1) \rceil$. Hence the procedure f-MATCHING runs in time $\mathcal{O}(\log^3 n)$. The procedure MODIFY consists only of simple computations on the adjacency lists. It takes $\mathcal{O}(\log n)$ times. Let us observe that the cost of each vertex cover in the input graph G is bounded by n^2 from above. Taking Lemma 11.3.5 into account we infer that the number of iterations of the **while** loop of the algorithm is $\mathcal{O}(\log n)$. Hence the algorithm runs in time $\mathcal{O}(\log^4 n)$.

□

11.4 Bibliographic notes

The *RNC*-algorithm for maximum f-matchings is from Garrido , Jarominek, Lingas and W. Rytter [GJLR96] and the deterministic *NC*-algorithm is from Diks, Garrido and Lingas [DGL93].

12

Parallelization of sequential algorithms

In this chapter we show several implementations of sequential algorithms. Unfortunately in the case of parallel matchings such an approach does not usually give NC-algorithms for general graphs, though a reasonable parallel speed-up can be achieved. In some special cases (e.g. expander graphs, approximations) the implementation of sequential algorithms leads to NC-algorithms.

We have already shown an application of *maximal* matchings to *maximum* matchings in the construction of an NC-algorithm for maximum matchings in dense graphs. In this chapter we show yet another application of NC-algorithms for maximal matchings: the sublinear time computation of maximum matchings in bipartite graphs.

Denote by NC_Maximal_Match(G) an NC-function returning any maximal matching of a given graph G. Any of the algorithms presented in Chapter 7 can be used as a function NC_Maximal_Match(G) returning a maximal matching and working in polylogarithmic time with a linear number of processors. A parallel version of Hopcroft–Karp algorithm will be presented using NC_Maximal_Match as the main subroutine in the crucial operation Disjoint_Paths.

Let us introduce the notation $\tilde{O}(f(n))$ for the asymptotic growth $O(f(n) \cdot \log^k(n))$, where k is a constant. In other words, in this chapter we neglect polylogarithmic factors, as the main part of the complexity will be in a form $O(n^\alpha)$, and the exponent α will be of main concern.

12.1 Parallelizing the Hopcroft–Karp algorithm

In the algorithm of Hopcroft and Karp a current matching M can be augmented in one stage *in parallel along all* paths in *any* maximal (with respect to inclusion) set Π = Disjoint_Paths of disjoint augmenting paths of (the same) length. Let us call such procedure Parallel_Augment(Π, M). It is easy to see the following:

Observation If Π is a set of vertex disjoint augmenting paths then the operation Parallel_Augment(Π, M) can be performed in $O(\log(n))$ time with a linear number of processors.

For a bipartite graph G and a matching M construct the graph Digraph(G, M) which is a directed version of G: the unmatched edges are directed left to right, and matched edges are directed right to left.

Let G' be a directed graph. We construct a function Parallel_BFS(A_0, B_0, G', k_0) which returns a distance from the set A_0 to B_0 in the graph G' and the *layer graph* L consisting of layers $0, 1, \ldots, k$, such that L_i is the set of nodes at distance i from A_0 for $i < k$ and L_k is the set of nodes in B_0 at distance k from A_0. If the distance exceeds k_0 the function returns $k = \infty$ and the empty layer graph. It can be interpreted as a *depth-restricted BFS*.

An example graph, its directed version $\vec{G}$ = Digraph(G) and the layer graph L constructed by Parallel_BFS($A_0, B_0, \vec{G}, k_0$) are illustrated in Figure 12.1.

Denote by Π = Disjoint_Paths(L, k) any maximal set of vertex-disjoint paths from A_0 to B_0 of length k in the layer graph L. The set Π is illustrated in Figure 12.2 together with the next matching which is the result of applying Parallel_Augment(Π, M) to the matching M from Figure 12.1.

```
Algorithm Parallel_Hopcroft_Karp;
{G is a bipartite graph (A, B, E), k0 will be determined later}
M := ∅; k :=0;
PHASE1: {short augmenting paths}
  repeat forever
    G⃗ := Digraph(G, M); {A0, B0 are the actual sets of free vertices in A, B};
    (L, k) := Parallel_BFS(A0, B0, G⃗, k0);
    {it is possible that k = ∞: no augmenting path}

    if k > k0 {there is no short augmenting path} then goto PHASE2;

    Π := Disjoint_Paths(L, k); M := Parallel_Augment(Π, M).
PHASE2: {long augmenting paths}
  repeat forever
    π := Single_Path(A0, B0); M := Augment(π, M)
    if π = nil {there is no augmenting path} then return M and STOP;
```

Lemma 12.1.1
The Hopcroft–Karp algorithm makes at most k_0 iterations in PHASE1 *and at most n/k_0 iterations in* PHASE2.

Proof A trivial consequence of Lemma 2.1.2 and Theorem 2.1.4 from Chapter 2.

□

12.2 Parallel implementation of BFS and Disjoint_Paths

In this section we specify the implementation details for the algorithm Parallel_Hopcroft_Karp, and we analyze the complexity of three basic functions in the algorithm: Parallel_BFS, Disjoint_Paths, Single_Path (the most important is the second one).

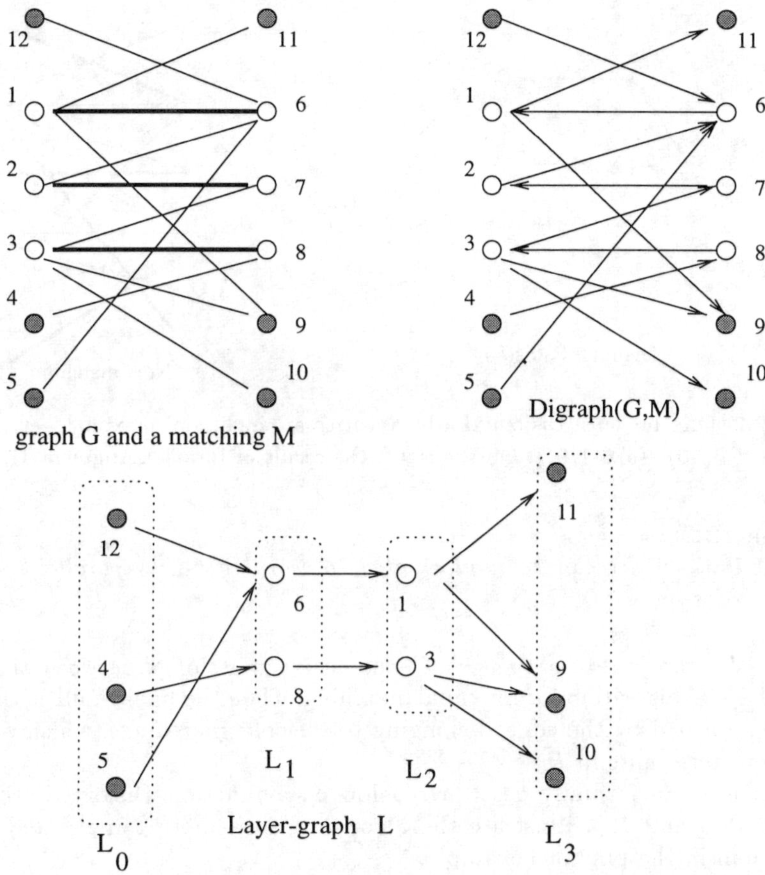

Fig. 12.1 A graph G with an initial matching (indicated in bold), Digraph(G, M) and the layer graph.

Lemma 12.2.1
Parallel_BFS$(A_0, B_0, \vec{G}, k_0)$ *can be implemented to work in* $\tilde{O}(k_0)$ *parallel time with* $O(n^2)$ *processors.*

Proof We produce the sequence of consecutive disjoint *layers*

$$A_0 = L_0, L_1, L_2, \ldots, L_{k_0} \subseteq B_0,$$

where, for $1 \leq i < k$, L_i is the set of vertices at distance 1 from L_{i-1} which are not in $L_0 \cup L_1 \cup \ldots \cup \mathrm{L}_{i-1}$. L_{k_0} is the set of nodes in B_0 at distance k from A_0.

We perform k *large* parallel steps; in one such step we compute the next layer. In each step we examine in parallel all edges of the graph, so a quadratic number of processors is sufficient.

□

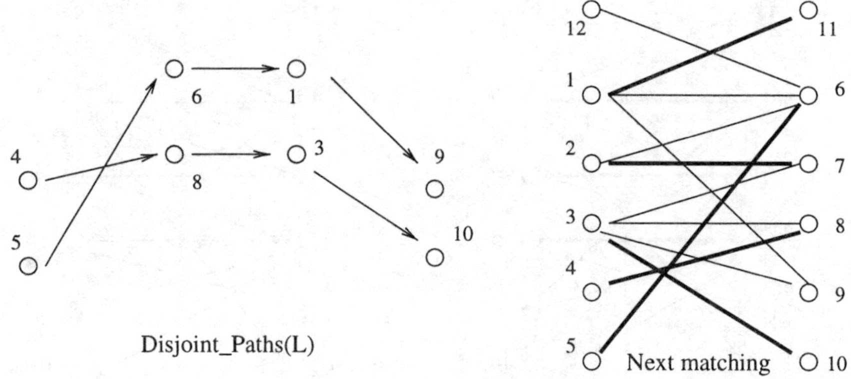

Fig. 12.2 On the left: Disjoint_Paths returns an inclusion maximal set of paths $\Pi = \{(4, 8, 3, 10),\ (5, 6, 1, 9)\}$. On the right, the result of Parallel_Augment(Π, M).

Lemma 12.2.2
Disjoint_Paths(L, k) *can be implemented to work in* $\tilde{O}(k_0)$ *parallel time with* $O(n^2)$ *processors.*

Proof We compute a sequence of consecutive matchings between the layers using the NC-algorithm for maximal matching. Then the final result is the set of all paths formed by the edges belonging to selected matchings, which originate in A_0 and terminate in B_0.

This is more formally described below as a function Disjoint_Paths(L, k). Figures 12.3 and 12.4 illustrate the algorithm on a more complicated layered graph than in the previous example.

□

We omit a simple proof of the following lemma. The underlying algorithm consists essentially in the computation of a transitive closure by squaring the adjacency matrix of the graph.

Lemma 12.2.3
Single_Path(A_0, B_0) *can be computed in* $O(\log^2 n)$ *time with a cubic number of processors.*

Theorem 12.2.4
There is an $\tilde{O}(n^{2/3})$ *parallel time algorithm for maximum bipartite matchings, the algorithm using* $O(n^3)$ *processors.*

Proof Take $k_0 = n^{1/3}$. Then the thesis is the direct consequence of Lemma 12.1.1, Lemma 12.2.1, Lemma 12.2.2 and Lemma 12.2.3.

□

```
function Disjoint_Paths(L, k);
{L_0, L_1, L_2, ..., L_k are the layers of L}  S := L_0;
for i := 1 to k do
    begin
        M_i := a maximal matching between S and L_i;
        {use NC_Maximal_Match}
        S := set of endpoints of M_i in L_i;
    end

return all paths consisting of edges in M_1 ∪ M_2 ∪ ... ∪ M_k
        starting in L_0 and terminating in L_k;
```

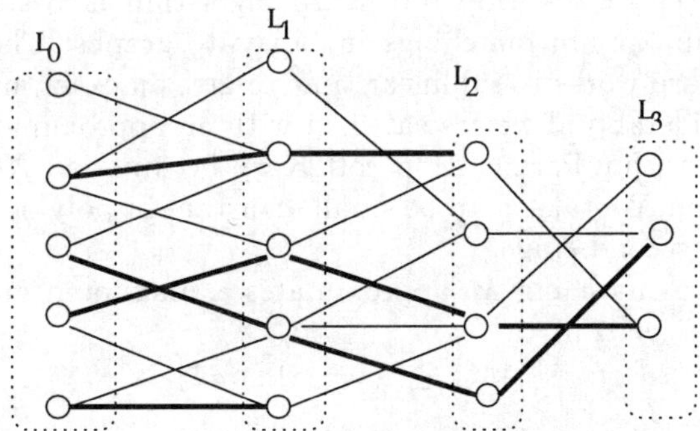

Fig. 12.3 Another example of a layered graph L; the matchings M_1, M_2, M_3 selected in the function Disjoint-Paths correspond to bold edges.

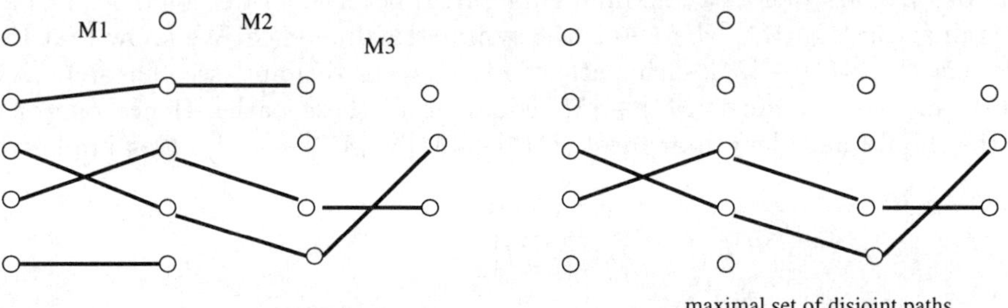

Fig. 12.4 On the left, the selected matchings form a set of paths originating in A_0. On the right, Disjoint_Paths returns the set of those paths which terminate in B_0.

Theorem 12.2.4 can be extended to weighted bipartite graphs using parallelization of some algorithms for optimal flows in networks.

Theorem 12.2.5 ([GPV93])
The maximum weighted matching in bipartite graphs with n vertices can be computed in $\tilde{O}(n^{2/3})$ parallel time with a polynomial number of processors.

A stronger theorem was proven in [GPST92] for sparse graphs (having a small number of edges) using interior point methods from optimization theory.

Theorem 12.2.6 ([GPST92])
The maximum weighted matching in bipartite graphs with m edges can be computed in $\tilde{O}(m^{1/2})$ parallel time with a polynomial number of processors.

Unfortunately for dense graphs (which have $O(n^2)$ edges) this gives $\tilde{O}(n)$ parallel time, which is much worse than $\tilde{O}(n^{2/3})$ in Theorem 12.2.4.

12.3 NC-approximation of maximum matching

The *parallel power* of the Hopcroft–Karp algorithm is best seen in the approximation of maximum matchings in bipartite graphs. The algorithm Parallel_Hopcroft_Karp works in sublinear time to give an exact answer, but we can speed it up considerably if we are satisfied with an approximate answer. In this situation the algorithm Parallel_Hopcroft_Karp becomes an NC-algorithm by a slight modification. We take k_0 to be small (constant or polylogarithmic at most) and we drop the second stage.

We say that a matching M approximates a maximum matching M^* with *approximation factor* α iff

$$|M| \geq (1-\alpha) \cdot |M^*|.$$

We now strenthen slightly the second point of Theorem 2.1.1.

Lemma 12.3.1
Assume the number of edges of a shortest augmenting path w.r.t. a matching M equals $2k+1$. Then M approximates the maximum matching with factor $1/k$.

Proof Let M^* denote a maximum matching. Let us consider augmenting paths contained in $M \oplus M^*$, where $\oplus$ is the symmetric difference. We know that there are exactly $|M^*| - |M|$ such paths and they are disjoint, see Theorem 2.1.1. There are also at most $|M^*| + |M|$ edges in all these paths. Hence one of the paths should have length at most $(|M^*| + |M|)/(|M^*| - |M|)$. This implies

$$\frac{|M^*| + |M|}{M^*| - |M|} \geq 2k+1.$$

It is matter of simple calculation to derive from the above inequality that $|M| \geq |M^*| \cdot (1 - 1/k)$.

□

Theorem 12.3.2 (approximation theorem for bipartite graphs)
Assume r is a constant. Then there is an NC-algorithm computing a matching in bipartite graphs with approximation factor $1/\log^k n$.

Proof Take $k_0 = 2 \cdot \log^k n + 1$.

The algorithm Parallel_Hopcroft_Karp is modified. The second stage is omitted. The algorithm stops and returns the actually computed matching in the moment when it tries to go to the second stage or stops. In this moment the shortest augmenting path exceeds k_0 or there is no such path at all (M is maximum).

Lemma 12.3.1 guarantees that the modified algorithm Parallel_Hopcroft_Karp approximates the maximum matching with factor $1/\log^k n$.

□

A similar approach, based on Lemma 12.3.1, was used in [FGHP93] for general graphs. However it does not give an *NC* approximation algorithm with ratio $1/\log^k n$. The algorithm of [FGHP93] is rather a *brute-force* algorithm and the degree of the polynomial describing the number of processors depends strongly on the approximation ratio $1/c$.

Theorem 12.3.3 (approximation theorem for general graphs)
Assume c is a constant. Then there is an NC-algorithm computing a matching in bipartite graphs with approximation factor $1/c$.

Proof Take $k_0 = 2c + 1$. We can consider the set $\mathcal{S}_i$ of all possible augmenting paths of a given constant length $i \le k_0$. It is crucial, since k_0 is constant, that the number of all such paths is polynomial. Hence we can create $\mathcal{S}_i$ and manipulate its elements in *NC*. Then, using the parallel algorithm for maximal independent set, we can choose a maximal set of disjoint augmenting paths and proceed in a similar way as in the first phase of the algorithm Parallel_Hopcroft_Karp. The resulting algorithm FGHP is presented below.

□

```
Algorithm FGHP;
{of Fisher, Goldberg, Haglin and Plotkin}
M := ∅;

for i := 1 to k0 do
    begin
      Si := set of all augmenting paths of length i;
      Gi := intersection graph for Si;
      Π := maximal independent set in Gi;
      {a maximal set of disjoint augmenting paths of length i}
      M := Parallel_Augment(Π, M);
    end
return M
```

12.4 Bipartite expander graphs

Bipartite expander graphs form a large class of important graphs for which algorithm Parallel_Hopcroft_Karp works in polylogarithmic time. This happens owing to the special *small distance property*:

> if a matching M is not a maximum cardinality matching then the length of a shortest augmenting path is bounded from above by $c \cdot \log n$, for a constant c.

The small distance property guarantees that in the Parallel_Hopcroft_Karp algorithm we can take $k_0 = c \cdot \log n$ and PHASE2 can be omitted.

The problem of maximum matching in bipartite expander graphs has many applications in sorting networks, permutation networks, path selection and routing in networks. Expander graphs are also part of the design of other special graphs: namely, concentrators and superconcentrators.

Let $G = (X, Y, E)$ be a bipartite graph with bipartition (X, Y), where $|X| = |Y| = n$. Denote by $N(A)$ the set of neighbors of a set of vertices A. Let $\beta > 1$ be a constant. We say that G is a left-sided *expander graph* with expansion β iff for every subset $A \subseteq X$ we have

$$|N(A)| \geq \min\{\beta \cdot |A|, n\}.$$

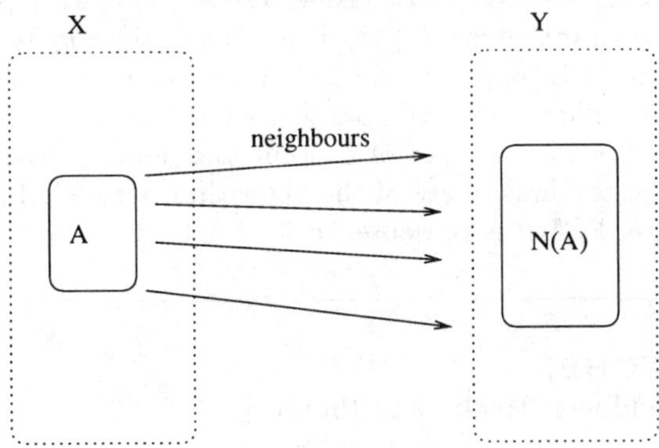

Fig. 12.5 Expanding property.

Lemma 12.4.1
Bipartite expander graphs have the small distance property. Assume l is the length of a shortest augmenting path with respect to a nonmaximum matching M; then $l \leq 2 \cdot \log_\beta n + 1$.

Proof Let U_0 be the set of *free* vertices in X and let U_k be the set of vertices in X accessible from U_0 by alternating paths of length at most $2k$. Then the β-expansion property easily implies the following

Claim Assume $2k < l$. Then $|U_{k+1}| \geq \beta \cdot |U_k|$.

Now take $k_{\max} = (l-1)/2$. The claim implies the following inequality

$$\beta^{(l-1)/2} \leq |U_{k_{\max}}| \leq n.$$

Consequently $l \leq 2 \cdot \log_\beta n + 1$.

□

The direct consequence of the last two lemmas is the existence of an *NC*-algorithm for maximum matchings in bipartite expander graphs.

Theorem 12.4.2
There is an NC-algorithm for maximum matchings in bipartite expander graphs.

Proof Take $k_0 = \log_\beta n$ and remove PHASE2 in the Parallel_Hopcroft_Karp algorithm.

□

12.5 Applying the planar separator theorem

The maximum matching in planar graphs can be made sequentially faster using a very strong property of planar graphs related to small separators.

Theorem 12.5.1 (planar separator theorem, [LT79], [KR87], [GM87])
Assume $G = (V, E)$ is a planar graph. Then there is a set $S \subseteq V$ satisfying:

1. $|S| = O(\sqrt{|V|})$.
2. *for each connected component C_i of $G - S$, $|C_i| \leq \frac{2}{3}|V|$.*
3. *S can be found in polylogarithmic time with $\tilde{O}(n^{1.5})$ processors.*

The set S from the theorem is called a *small planar separator.*

Denote by MaxCardMatch(G) a maximum cardinality matching of a graph G.

The algorithm for maximum matching in planar graphs is based on the fact that we can compute independently maximum matchings M_i in the components C_i obtaining maximum matching in the graph $G - S$. Then we have a good approximation of a globally maximum matching due to Lemma 12.5.2; it is also illustrated in Figure 12.6.

The structure of the algorithm is also similar to the Parallel_Hopcroft_Karp algorithm and consists of two phases.

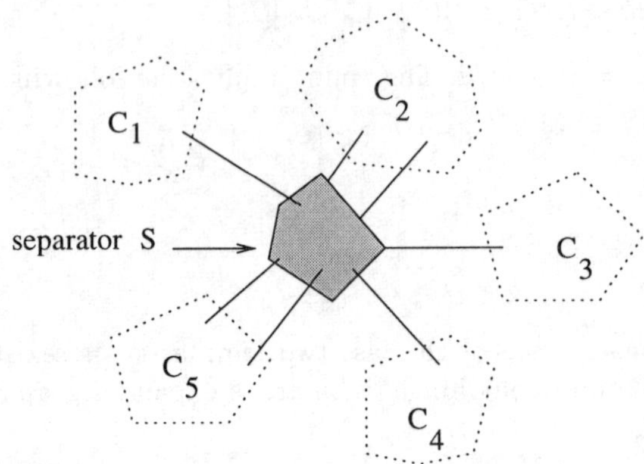

Fig. 12.6 A small separator S in a planar graph. If M_i is a maximum matching in C_i and M is a maximum matching in G then $|M| - \sum_i |M_i| \le |S|$.

```
function MaxCardMatch(G);
PHASE1: {approximate maximum matching}
  S := small separator of G;
  compute connected components C_1, C_2, ..., C_k of G - S;
  for each 1 <= i <= k do in parallel
    M_i := MaxCardMatch(C_i);

M := ∪ M_i;
PHASE2: {additional O(√|V|) augmentations}
  repeat forever
    G⃗ := Digraph(G, M); {A_0, B_0 are the actual sets of free vertices in A, B};
    π := Single_Path(A_0, B_0); M := Augment(π, M)
    if π = nil {there is no augmenting path} then return M and STOP;
```

Lemma 12.5.2
Assume S is a small planar separator of G. M = MaxCardMatch(G) *and M' =* MaxCardMatch($G - S$). *Then*

$$|M - M'| = (\sqrt{|V|}).$$

Proof If we remove the edges in M incident to nodes in S then we receive a matching M'' in $G - S$. However, we removed only $O(\sqrt{|V|})$ since S is a small separator. Now the thesis follows from the fact that $|M'| \ge |M''|$.

□

Lemma 12.5.3 ([K86])
The function Single_Path *in planar graphs can be implemented to work in polylogarithmic time with $\tilde{O}(n^{1.5})$ processors, and can be found with $\tilde{O}(n^{1+\epsilon})$ in the case of randomized algorithms.*

Theorem 12.5.4 ([K86], [DKL89])
A maximum matching in planar bipartite graphs can be found deterministically in $\tilde{O}(\sqrt{n})$ parallel time using $\tilde{O}(n^{1.5})$ processors. The number of processors can be reduced to $\tilde{O}(n^{1+\epsilon})$ in the case of randomized algorithms.

Proof The number of processors is as required by Lemma 12.5.3 and due to the planar separator theorem. The randomized processor-efficient algorithm for planar separators was given in [GM87].

If $T(n)$ is the time complexity of the algorithm, then the *divide-and-conquer* structure of the algorithm implies the following recurrence:

$$T(n) \leq T(\tfrac{2}{3}n + \tilde{O}(\sqrt{n}).$$

It is easy to estimate $T(n) = \tilde{O}(\sqrt{n})$.

□

Our sublinear time algorithm is very simple. It is also possible to have a more complicated algorithm which reduces the planar bipartite matching problem to multi-source multi-sink flows in a planar graph. The latter problem is in *NC*, see [MN95]. In the bipartite graph $G = (X, Y, E)$ all nodes of X become sources and all nodes of Y become sinks. The edges are directed from X to Y. The demand at each source and sink is one unit, and the edge capacities are also one unit each, see Figure 12.7.

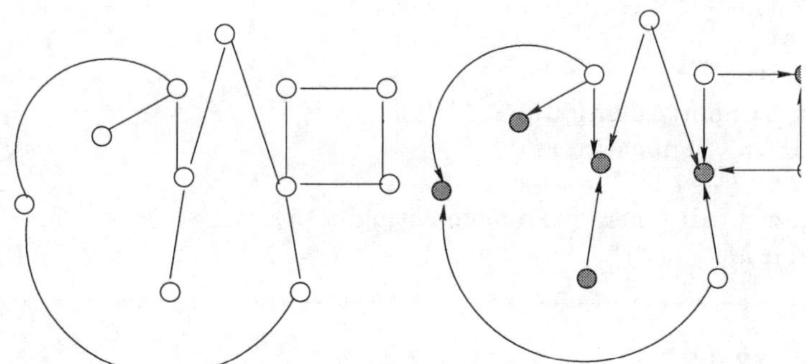

Fig. 12.7 Changing a bipartite planar graph into a planar multi-source multi-sink network: the sinks are shaded.

This gives the following result as an application of flows in planar networks; we omit the proof.

Theorem 12.5.5 ([MN95], [J87])
The perfect matching problem in planar bipartite graphs is in NC.

12.6 Parallel version of the Edmonds algorithm

The book is devoted to *fast* parallel algorithms, which means mostly polylogarithmic (or, eventually, sublinear time) algorithms. However, in this section we also mention an interesting $\tilde{O}(n)$ parallel time algorithm. We omit the details as these are not the main focus of the book, but refer the reader to [SM86] for them. The algorithm of Schieber and Moran works for a maximum matching problem, but for simplicity we consider only the perfect matching testing problem (observe that the maximum matching is polylogarithmically reducible to the latter problem).

Assume we have a function ShortAugmPath(G, M, k) which returns an augmenting path w.r.t. M of length at most k. If there is no such path then the function returns *nil*.

Lemma 12.6.1
The value of ShortAugmPath(G, M, k) *can be computed in* $\tilde{O}(k)$ *time with a polynomial number of processors.*

Proof The details of a rather technical proof are omitted. A kind of parallel *BFS* is applied, which is an implementation of the Edmonds algorithm but the explicit shrinking of blossoms is avoided. We refer to [SM86] for details. □

Algorithm SM {of Schieber and Moran};
{our version tests only for a perfect matching}
$M := \emptyset$; $k := 0$;
repeat
  $l_k := \frac{n}{n/2-k}$;
  $\pi :=$ ShortAugmPath(G, M, l_k);
  $M :=$ Augment(π, M);
  $k := k + 1$;
until $\pi =$ **nil** {there is no augmenting path};
return M;

Theorem 12.6.2
The perfect matching problem in general graphs can be solved in $\tilde{O}(n)$ *parallel time with a polynomial number of processors.*

Proof After k iterations of the algorithm SM the actual matching has size i. Assume that the maximum cardinality matching $M^*|$ has size $n/2$. Then, from Theorem 2.1.1 (see Chapter 2), the length of a shortest path is bounded by

$$l_k = \frac{n}{n/2 - k}.$$

Hence, by Lemma 12.6.1, the total time of the algorithm is bounded by

$$\tilde{O}(\frac{n}{n/2} + \frac{n}{n/2-1} + \frac{n}{n/2-2} + \ldots + \frac{n}{n/2-(n/2-1)}) = \tilde{O}(n \cdot \sum_{k=1}^{n} \frac{1}{k}) = \tilde{O}(n).$$

□

An interesting open question is to design a *deterministic sublinear* time parallel algorithm working with a polynomial number of processors for the matching problem in *general graphs.*

12.7 Bibliographic notes

The algorithm working in $O(n^{2/3})$ time for weighted bipartite matchings was constructed in Goldberg, Plotkin, and Vaidya [GPV93]. It was improved for sparse graphs in Goldberg, Plotkin, Shmoys and Tardos [GPST92]. Our exposition of matchings in bipartite graphs is based on Haglin [H].

The efficient sublinear time algorithms for bipartite planar graphs are from Kasteleyn [K67] and Dahlhaus, Karpinski and Lingas [DKL89]. The approximation algorithm for general graphs is based on Fisher, Goldberg, Haglin and Plotkin [FGHP93]. The $\tilde{O}(n)$ time deterministic algorithm for general graphs is from Schieber and Moran [SM86].

13

Pfaffians, counting the number of matchings, and planar graphs

In this chapter we study the notion of the Pfaffian of a graph and its connection to the problem of counting the number of matchings, and the problem of computing a *Pfaffian orientation of a planar graph.*

13.1 Pfaffians of graphs

In Section 6.1 we introduced the notions of Edmonds' and Tutte's matrices, and investigated the relation of their determinants to the problem of deciding the existence of a perfect matching. Towards the end of this section we will introduce the notion of a *Pfaffian* of Tutte's matrix A corresponding to the two-cycle permutations over $\{1, \ldots, 2n\}$.

Given an arbitrary graph $G = (V, E)$ with $|V| = n$, an *orientation* (graph) of G is any directed graph $\vec{G} = (V, \vec{E})$, $E \subseteq V \times V$, such that $<u, v> \in \vec{E}$ if and only if $(u, v) \in E$ and $<v, u> \notin \vec{E}$. We construct an $n \times n$ *skew adjacency matrix* $S_{\vec{G}} = [s_{ij}]$ of $\vec{G}$ as follows:

$$s_{ij} = \begin{cases} 1 & \text{if } <u_i, u_j> \in \vec{G}, \\ -1 & \text{if } <u_j, u_i> \in \vec{G}, \\ 0 & \text{elsewhere.} \end{cases}$$

Observe that there is a one-to-one correspondence between the perfect matchings of G and the nonzero monomials of $\text{PFAFFIAN}(S_{\vec{G}})$. Indeed, for $M(G)$, the number of perfect matchings in G, we have

$$|\text{PFAFFIAN}(S_{\vec{G}})| \leq M(G).$$

If the above inequality becomes the equality we call the directed graph $\vec{G}$ *Pfaffian* (or the *Pfaffian orientation*). If for a given graph $G = (V, E)$ such a Pfaffian orientation exists, we call G *Pfaffian.*

Now we pause for a moment to establish an important connection between Pfaffians and the determinants of general skew-symmetric matrices.

13.2 Skew-symmetric determinants and Pfaffians

In this section we prove rigorously an important fact behind Lemma 6.2.4 that for any skew-symmetric matrix A (a matrix A such that its *transpose* $A^\top = -A$) over an arbitrary field of characteristic $\neq 2$, $\det(A) = (\text{PFAFFIAN}(A))^2$. A determinant of a skew-symmetric matrix is also called *skew-symmetric*.

We shall now consider a matrix A over an arbitrary field of characteristic $\neq 2$.

Lemma 13.2.1
Given any $n \times n$ skew-symmetric matrix A for n odd, then $\det(A) = 0$.

Proof $\det(A) = \det(A^\top) = \det(-A) = (-1)^n \det(A)$.

□

Lemma 13.2.2
Given any $n \times n$ skew symmetric matrix A for n even, then

$$\det(A) = (\text{PFAFFIAN}(A))^2.$$

Proof Let $A = [a_{ij}]$ be skew symmetric. We note also that the matrix $A' = B^\top AB$, for an arbitrary $n \times n$ matrix B, is skew symmetric. We have $A'^\top = (B^\top AB)^\top = B^\top A^\top B = -B^\top AB = -A'$. We shall prove our lemma by induction on $n = 2k$. For $k = 1$ we have $\det(A) = a_{12}^2$ and $\text{PFAFFIAN}(A) = -a_{12}$. Suppose now that $\det(A) = (\text{PFAFFIAN}(A))^2$ for all $2k \times 2k$ matrices $A = [a_{ij}]$. We prove that this is also true for all the $(2k+2) \times (2k+2)$ skew-symmetric matrices C.

Define $C_{2k} = [c_{ij}]$ to be a $2k \times 2k$ skew-symmetric matrix with $k \times 2k$ variables x_{ik}, such that $c_{ik} = x_{ik}$ for $i < k$ and $c_{ki} = -c_{ik}$ and $c_{ii} = 0$, elsewhere. Construct a $(2k+2) \times (2k+2)$ matrix $B = [b_{ij}]$ such that $b_{ii} = 1$ and $b_{ij} = 0$ for $j \neq 1$ and $i \neq j$, and

$$b_{i1} = d_i = -\frac{c_{i,2n+2}}{c_{1,2n+2}} \quad \text{for} \quad i \neq 1;$$

$$B = \begin{bmatrix} 1 & 0 & 0 & \dots & 0 \\ d_2 & 1 & 0 & \dots & 0 \\ d_3 & 0 & 1 & \dots & 0 \\ \vdots & \vdots & \vdots & \ddots & \vdots \\ d_{2n} & 0 & 0 & \dots & 1 \end{bmatrix}.$$

Note that $\det(B) = 1$, and

$$BC_{2k+2}B^\top = \begin{bmatrix} 0 & \dots & x_{1,2n+2} \\ \vdots & & 0 \\ \vdots & C_{2k} & \vdots \\ 0 & & \vdots \\ -x_{1,2n+2} & \dots & 0 \end{bmatrix}.$$

We have

$$\begin{aligned}\det(BC_{2k+2}B^{\intercal}) &= (\det(B))^2 \det(C_{2k+2}) \\ &= \det(C_{2k+2}) \\ &= (x_{1,2n+2})^2 \det(C_{2k}).\end{aligned}$$

By the induction hypothesis $\det(C_{2k}) = (\text{PFAFFIAN}(C_{2k}))^2$, and therefore $\det(C_{2k+2}) = (x_{1,2k+2})^2 \cdot (\text{PFAFFIAN}(C_{2k}))^2 = (\text{PFAFFIAN}(C_{2k+2}))^2$, and our lemma follows.

□

Lemmas 13.2.1 and 13.2.2 entail the following:

Corollary 13.2.3
Given any skew symmetric matrix A, the following identity holds:

$$\det(A) = (\text{PFAFFIAN}(A))^2.$$

This gives the proof of Lemma 6.2.4 of Chapter 6.

13.3 Pfaffians and counting the number of perfect matchings

Given a graph $G = (V, E)$, let $\vec{G} = (V, \vec{E})$ be any orientation of G, and C be any (undirected) even cycle in G. Observe that for any *routing* of C, if C contains an even number of oriented edges of $\vec{G}$ with orientation agreeing with this routing, then C also contains an even number of edges with orientation contrary to the routing. Therefore the following definition is independent of the chosen routing.

Definition 13.3.1
Let $\vec{G}$ be any orientation of a graph G, and C any (undirected) even cycle in $\vec{G}$. We say C is evenly oriented if it has an even number of edges with orientation agreeing with the routing. Otherwise, we say C is oddly oriented.

For a graph G and a perfect matching M of G define sign(M) *to be equal to the* sign *of a corresponding monomial in its Pfaffian.*

We formulate without a proof (cf., e.g. [LP86]) the following:

Lemma 13.3.2
Given an arbitrary graph G, and $\vec{G}$ an arbitrary orientation of G, let M_1 and M_2 be any two perfect matchings of G and let N denote the number of evenly oriented alternating cycles in $M_1 \cup M_2$. Then

$$\text{sign}(M_1) \cdot \text{sign}(M_2) = (-1)^N.$$

For a given cycle C of a graph $G = (V, E)$ denote by $V(C)$ the set of all vertices in C. We call C *nice* if there exists a perfect matching in C restricted to $V \setminus V(C)$.

The last inequality of Section 13.1 and Lemma 13.3.1 entail the following (cf. [LP86]):

Theorem 13.3.3
Let $G = (V, E)$ be an even graph ($|V|$ is even), and $\vec{G}$ any orientation of G. The following properties are equivalent:

(i) Every perfect matching of G has the same sign *relative to $\vec{G}$.*

(ii) $\vec{G}$ is a Pfaffian orientation of G.

(iii) Every nice cycle in G is oddly oriented relative to $\vec{G}$.

The above suggests the method of counting the number of perfect matchings in Pfaffian graphs $\vec{G}$: construct a Pfaffian orientation $\vec{G}$ in a Pfaffian graph and its skew adjacency matrix $S_{\vec{G}}$. By Corollary 13.2.3, the number of perfect matchings $M(G)$ of G is equal to $\sqrt{\det(S_{\vec{G}})}$.

The important classes of graphs for which Pfaffian orientations can be constructed efficiently are planar graphs (cf. [K86], [K85] and [LP86]), or even a wider class of $K_{3,3}$-free graphs (cf. [L74]).

We shall now describe an efficient parallel algorithm for computing the number of perfect matchings in an arbitrary planar graph which is based on Kasteleyn's result (cf. [K86], [K85]) on computing Pfaffian orientations in this class of graphs.

We call a graph *plane* if it can be drawn on a plane in such way that each vertex is represented by a point, each edge by a continuous line connecting two points, and no two lines representing the edges intersect except their endpoints. If a graph G has such a *planar representation* $\overline{G}$ on the plane we call it planar. $\overline{G}$ will be called a planar representation of G. Planar graphs are interesting combinatorial objects with a number of direct practical applications especially in circuit and network design.

A planar representation $\overline{G}$ of a graph G divides the plane into connected regions, called *faces* of $\overline{G}$, each region bounded by lines representing edges (a boundary). One of the faces is of infinite area. It is called an *exterior* (or *infinite*) *face*.

We consider a connected plane graph G, that is a planar graph in which every two vertices are connected by a path. Suppose now that $\overline{G}$ is its planar representation with F a face. Suppose also that a line ℓ belongs to the boundary of F. We say that ℓ is *clockwise* oriented if F lies to the right of ℓ when proceeding along its direction, see Figure 13.1

Let us now orient $\vec{G}$ arbitrarily, and let B be a boundary of some face F of $\overline{G}$. The *orientation parity* of B is the parity of the number of lines in B which are oriented clockwise. We shall denote an oriented planar representation of G by $\vec{G}$. For a given cycle C of G the *interior* of C (with respect to the planar representation of G) is the set of all vertices lying in the interior of the corresponding face of $\overline{G}$.

Theorem 13.3.4 ([K86])
Let $\vec{G}$ be an oriented planar representation of a connected graph G such that every boundary of its finite (area) faces has the odd orientation parity. Then

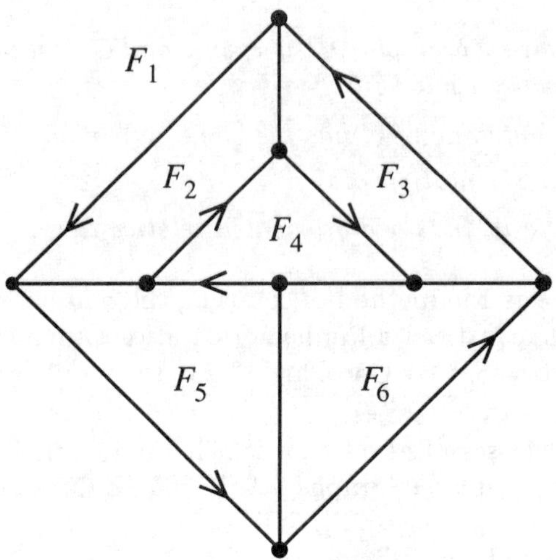

Fig. 13.1 Faces of $\overline{G}$, and the clockwise orientations of the boundaries of F_1, and F_4.

in every cycle of G the orientation parity of a corresponding boundary has the opposite parity to the number of vertices in the interior of a cycle.

Proof Let C be a cycle of $\vec{G}$, and there are K faces, $F_1, \ldots, F_K$, enclosed by C. Denote their corresponding boundaries by $B_1, \ldots, B_K$, and their respective orientation parities by $p_1, \ldots, p_K$. By hypothesis

$$K \equiv \sum_{i=1}^{K} p_i \pmod 2.$$

Let s be the number of vertices inside C, e the number of edges inside the interior of C, and m the length (number of edges) of C. By *Euler's formula* (cf., e.g. [Gi85], p.69) we have the following equality for the cycle C with its interior:

$$K + 1 = (e + m) - (s + m) + 2.$$

This implies that the number e of edges inside the interior of C satisfies

$$e = s + K - 1.$$

Denote by p the orientation parity of C. We then have the following equality:

$$\left(\sum_{i=1}^{K} p_i\right) \bmod 2 = (p + e) \bmod 2$$

as each interior line in the boundary of a given face gets a clockwise orientation only once.

Therefore $K \equiv p + e \pmod 2$, and consequently $K \equiv p + s + K - 1 \pmod 2$. We have $p + s - 1 \pmod 2$, and $p \equiv (s - 1) \pmod 2$. Therefore the orientation parity of the boundary corresponding to C has an opposite parity to the number of vertices in the interior of a cycle, and our theorem follows.

□

Theorems 13.3.2 and 13.3.3 now entail the following (we view $\vec{G}$ of Theorem 13.3.3 simply as a directed graph).

Corollary 13.3.5
The directed graph which corresponds to the oriented planar representation $\vec{G}$ of Theorem 14.3.3 is Pfaffian.

Proof If G is an odd graph any orientation $\vec{G}$ of G is Pfaffian. Suppose that G is an even graph. We take the oriented planar representation $\vec{G}$ from Theorem 13.3.3. Suppose that C is a nice cycle of G, that is there exists perfect matching M in $V \setminus V(C)$. This means the parity of $V \setminus V(C)$ is even, and therefore C must be oddly oriented relative to $\vec{G}$. Our corollary now follows from Theorem 14.3.2.

□

We show now how to use Theorem 13.3.3 for an efficient parallel construction of a Pfaffian orientation for any given connected planar graph $G = (V, E)$.

We use a simple inductive argument on $|E|$. We note also that if G is a tree any orientation will be Pfaffian. Suppose that G is not a tree, and choose any line e from the boundary B of the infinite face of its planar representation $\overline{G}$. Let B_1 be the boundary of the finite face, which boundary contains e. We construct the graph $\bar{G}_1$ by removing ℓ from $\overline{G}$. By the induction hypothesis $\bar{G}_1$ has an orientation $\vec{G}_1$ such that every boundary of a finite face of $\bar{G}_1$ has the odd orientation parity. We insert ℓ and orient it in such a way that the orientation parity of B_1 is odd. All the other boundaries different from B_1 will be unchanged, so the resulting orientation $\vec{G}$ of $\overline{G}$ will have the properties of Theorem 13.3.3.

Now the parallel (NC)-algorithm for computing a Pfaffian orientation of any connected planar graph is straightforward (cf. [V89]):

1. Construct a planar representation $\overline{G}$ of G using a planar embedding algorithm of [KR87].
2. Construct a spanning tree T of G (cf., e.g. [GR88]).
3. Construct the *faces-graph* H of $\overline{G}$ having vertices corresponding to the faces of $\overline{G}$. Two vertices of H are connected by an edge if the respective face boundaries share a line ℓ not in T. Such an edge is said to represent a line ℓ. The resulting graph is a tree. Pick up a vertex r in H which corresponds to the infinite face of $\overline{G}$, and root H at r.
4. Orient H arbitrarily, say all the edges directed away from the root r.
5. The tree H now includes the orientation $\vec{G}$ of $\overline{G}$ (and G). The orientation of lines in $\overline{G}$ starts with the boundaries of faces corresponding to the leaves of H. Let $e = (u, v)$ be an edge of G corresponding to the edge $\langle u, v \rangle$ of

H. Assume that the boundaries of faces corresponding to all descendants of v have already been oriented. Let F be the face of $\overline{G}$ corresponding to v. Then, e represents the only unoriented line ℓ in the boundary of F. Orient the line ℓ in such a way that the orientation parity is odd. This can be done in parallel maintaining the orientation parities of face boundaries in the vertices of H, and computing mod 2 sums of parities of v and all its descendants.

The above yields:

Theorem 13.3.6
Given any planar graph G, there exists an NC-algorithm for constructing its Pfaffian orientation $\vec{G}$.

We combine Theorem 13.3.5 and Corollary 13.2.3 to obtain

Theorem 13.3.7
Given any planar graph G, there exists an NC-algorithm for computing the number $M(G)$ of perfect matchings in G.

Proof By Corollary 13.2.3, $(\mathrm{PFAFFIAN}(S_{\vec{G}}))^2 = \det(S_{\vec{G}})$ for a Pfaffian orientation $\vec{G}$ of G (Theorem 13.3.5). We have $|\mathrm{PFAFFIAN}(S_{\vec{G}})| = M(G)$, and $\mathrm{PFAFFIAN}(S_{\vec{G}}) = \sqrt{\det(S_{\vec{G}})}$. Lemma 5.1.2 now entails our theorem. □

Theorem 13.3.6 can be generalized to hold for the $K_{3,3}$-free graphs (cf. [V89]) as well. It is still an open problem whether the construction of a perfect matching in a planar graph lies in *NC*. The status of this problem is very interesting in view of the corresponding status of its counting version proven above. The problem of constructing a perfect matching in a planar bipartite graph is known to be in *NC* ([MN95]).

13.4 Bibliographic notes

For an extensive treatment of Pfaffians, skew-symmetric matrices and computing Pfaffian orientations, see Muir [M05], Gröbner [G56], Lovász and Plummer [LP]. For efficient parallel algorithms on testing planarity, and the construction of planar representations of graphs, see Klein and Reif [KR87], and for maximum flow problems in planar graphs see Miller and Naor [MN95].

14

Basic applications of parallel matching

In this chapter we present some natural applications of parallel algorithms for *maximum* matchings. We show that due to RNC-algorithms for maximum matchings there exist RNC-algorithms for the following problems: maximum disjoint paths, optimal flows in some networks, DFS-tree construction and subtree isomorphism. If there is an NC-algorithm for maximum matchings then there are NC-algorithms for each of the above problems. Hence the maximum matching problem is the main representative of an important class of combinatorial problems NC-reducible to it.

14.1 The subtree isomorphism problem

Two unrooted undirected trees T', T'' are *isomorphic* (we write $T' \equiv T''$) iff they have *the same shape*, that is iff they are isomorphic in the sense of undirected unlabeled graphs. There are linear time sequential algorithms and optimal NC-algorithms to test tree isomorphism.

However, the problem of subtree isomorphism is much more complex. The subtree isomorphism problem can be defined as follows:

> there are two trees $T1$, $T2$, test if there is a subtree T of $T2$ such that $T1 \equiv T2$.

The main result of this section is an RNC-algorithm for the subtree isomorphism problem. We also show that this problem is in NC if and only if the perfect matching problem for bipartite graphs is in NC.

It is usually much easier to deal with *rooted directed trees*: two such trees T', T'' are isomorphic iff there is a bijection between their sets of nodes such that the roots correspond to each other, and the bijection preserves the relation "to be a father of".

The rooted subtree isomorphism problem is to test if a given rooted tree $T1$ is isomorphic to a directed subtree rooted at the root of $T2$ (if this happens then we write $T1 \preceq T2$).

Denote by $T(w)$ the directed subtree of T rooted at w. It is obvious that the subtree isomorphism is NC-reducible to the rooted subtree isomorphic (one has to consider all nodes u of $T2$ as possible roots and test $T1 \preceq T2(u)$, in parallel for all $u \in T2$).

Hence in this section we shall deal with the rooted directed subtree isomorphism problem.

We introduce the boolean table SUBTREE. For nodes $u \in T1$ and $w \in T2$

$$\text{SUBTREE}(u, w) = \textbf{true} \text{ iff } T1(u) \preceq T2(w).$$

It is clear that the subtree isomorphism problem is reducible to the computation of the table SUBTREE.

First we show how to reduce the perfect matching problem to the rooted subtree isomorphism problem. We define an easier version of the rooted subtree isomorphism. Consider two trees $T1$ and $T2$ rooted respectively at u and w. Assume for each son u_i of u and each son w_j of w we know if $T1(u_i) \preceq T2(w_j)$. The simple subtree isomorphism problem consists in testing if $T1 \preceq T2$. We denote this problem by MATCH_SUBTREES(u, w).

We shall also treat MATCH_SUBTREES as a function. If we know the values of the table SUBTREE(u_i, w_j) for all sons u_i of u and all sons w_j of w then MATCH_SUBTREES(u, w) returns SUBTREE(u, w).

The function MATCH_SUBTREES will be our main function in the parallel algorithm for the subtree isomorphism problem.

Assume for simplicity that u and w have the same number k of sons (we can add some dummy nodes if necessary). Define the bipartite graph $G(u, w) = (X, Y, E)$ such that

$$X = Y = \{1, 2, \ldots, k\}, (i, j) \in E \text{ iff } T1(u_i) \preceq T2(w_j).$$

The following fact follows directly from the definitions above.

Lemma 14.1.1
MATCH_SUBTREES(u, w) = *true iff the graph* $G(u, w)$ *has a perfect matching. The problem* MATCH_SUBTREES *is in RNC.*

Theorem 14.1.2
The perfect matching problem for bipartite graphs is NC-reducible to the rooted subtree isomorphism problem.

Proof For a given equibipartite graph $G = (X, Y, E)$ we can easily construct two trees $T1$ and $T2$ rooted at u and w, respectively, such that $G = G(u, w)$. Let $X = Y = \{1, 2, \ldots, k\}$. Assume for each i that subtree $T1(u_i)$ consists of a chain containing i nodes, see Figure 14.1.

Assume also that for each j the subtree $T2(w_j)$ consists of separate chains containing $i_1, i_2, \ldots, i_t$ nodes, where $i_1, i_2, \ldots, i_t$ are all neighbours (in X) of the j-th node in Y. Then $G(u, w) = G$ and the perfect matching problem for G is equivalent to the rooted subtree isomorphism problem for $T1$ and $T2$.

□

We now show how to make an NC-reduction of the rooted subtree problem to the computation of the function MATCH_SUBTREES. The computation of this

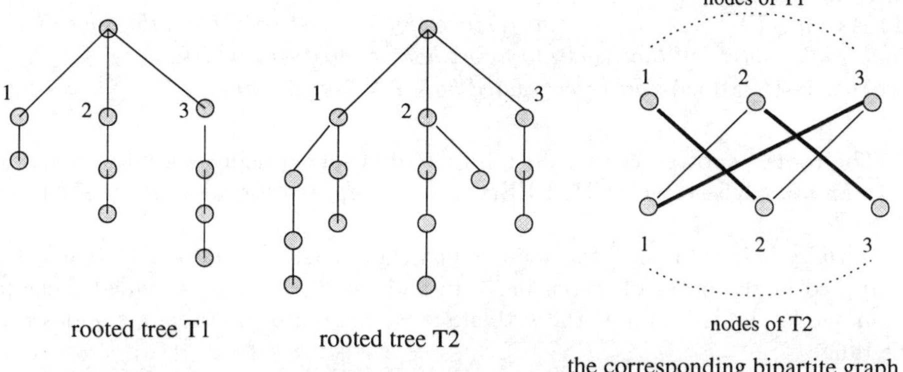

Fig. 14.1 The rooted subtree isomorphism problem for the tress $T1$ and $T2$ (on the left) is equivalent to the matching problem for the bipartite graph (on the right). There is an edge (i, j) iff $T1(i)$ is isomorphic to a subtree of $T2$ rooted at j.

function is the only place where we use the algorithm for the matching problem from Lemma 14.1.1.

We use some *tree-cutting* technique. Define the weight of a subtree as the number of its nodes, and define weight(v) as the weight of the subtree rooted at v.

For a give node its heaviest subtree is a subtree rooted at a son of this node and having maximum weight.

For a node $v \in T1$ denote by Heaviest_Path(v) the path π from v to a leaf; π starts at v and each time it goes down it descends to the heaviest subtree of a current node.

Two important properties of the heaviest paths are expressed by the lemma below. An easy proof is omitted.

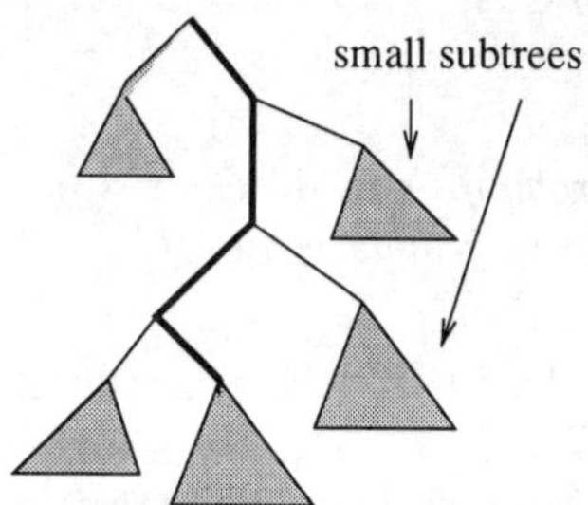

Fig. 14.2 A heaviest path (in bold). If n is the size (number of nodes) of the whole tree then the size of each shaded subtree is at most $\frac{n}{2}$.

Lemma 14.1.3
(1) *Assume $T1$ has m nodes and remove the heaviest path from the root of $T1$. Then each subtree of the resulting forest has at most $\frac{1}{2}m$ nodes.*
(2) Heaviest_Path(v) *can be computed by an NC-algorithm.*

The rooted subtree isomorphism problem is solved using a kind of *dynamic programming.* The table SUBTREE(u, w) is computed for all nodes $u \in T1$ and $w \in T2$.

Initially it is computed for nodes u of weight 1 (i.e. leaves of $T1$). Then it is computed in the order of increasing values of weight(u). The so-called *doubling technique* is applied, and at the k-th stage we compute the table for nodes u in the range

$$Range_k = \{\, u\colon\ 1 \le \text{weight}(u) \le 2^k \,\}.$$

Observation For $u \in Range_0$ (a leaf of $T1$) and any node $w \in T2$,

$$\text{SUBTREE}(u, w) = \textbf{true}.$$

For an edge $e = (x, y)$ of a tree T denote by Partial_Tree(e) the subtree $T(x)$ with the subtree $T(y)$ and the edge (x, y) removed.

Lemma 14.1.3 implies the following.

Lemma 14.1.4
Assume $v \in Range_k$ and $e \in$ Heaviest_Path(v). *Then all sons of the root of* Partial_Tree(e) *are in $Range_{k-1}$.*

Theorem 14.1.5
The subtree isomorphism problem is in RNC. It is in NC if and only if the perfect matching problem for bipartite graphs is in NC.

Proof The function MATCH_SUBTREES is reducible to the matching problem owing to Lemma 14.1.1.

The algorithm Rooted_Subtree_Isomorphism (presented below) is based on Lemma 14.1.4 and computes the table SUBTREE for nodes $u \in T1$ successively in $Range_0, Range_1, \ldots, Range_{\log|T1|}$. Figure 14.3 demonstrates the computation of SUBTREE(u, w). The algorithm performs a logarithmic number of iterations.

At the end we return the value of the entry in the table SUBTREE corresponding to the roots of the trees $T1$ and $T2$.

```
Algorithm Rooted_Subtree_Isomorphism(T1, T2);
{first computes SUBTREE(u, w) for all u ∈ T1, w ∈ T2}

initialization:
        for each u ∈ T1, w ∈ T2 do in parallel
            SUBTREE(u, w) := (u is a leaf in T1);

main iteration:
        for k := 1 to log_2 |T1| do
        for each u ∈ Range_k, w ∈ T2 do in parallel
        begin
            for each edge e = (x, y) of T2(w) do in parallel
            begin
            e' := HeaviestEdge(T1(u), depth_{T2(w)}(x));
            MATCH_EDGE_{u,w}(e) :=
                    MATCH_SUBTREES(Partial_Tree(e), Partial_Tree(e'));
            end ;
        if there is a path π down from w of length |Heaviest_Path(u)|
        such that for each e ∈ π MATCH_EDGE_{u,w}(e) = true
                    then SUBTREE(u, w) := true;
        end ;

if SUBTREE(root(T1), w) for some node w ∈ T2 then
return true {T1 is isomorphic to a rooted subtree of T2}
        else return false
```

Testing in a tree for a path of a given length (going down from a specified node) with all edges labeled **true** is easily computable by an NC-algorithm. Hence the subtree isomorphism is NC-reducible to the matching problem.

□

14.2 Maximum flows and disjoint paths

The *maximum flow* problem is one of the basic optimization problems. It is solvable sequentially in polynomial time, but it is hardly parallelizable because of the following fact.

Theorem 14.2.1 ([GR88])
The maximum flow problem is P-complete.

So we cannot expect that the problem is in NC or RNC, unless $P = NC$ or $P = RNC$, respectively. However, the proof of P-completeness requires exponential capacities (though their binary representation is linear). In this section we show that the same problem but with polynomial capacities is in RNC, since it is NC-reducible to the matching problem.

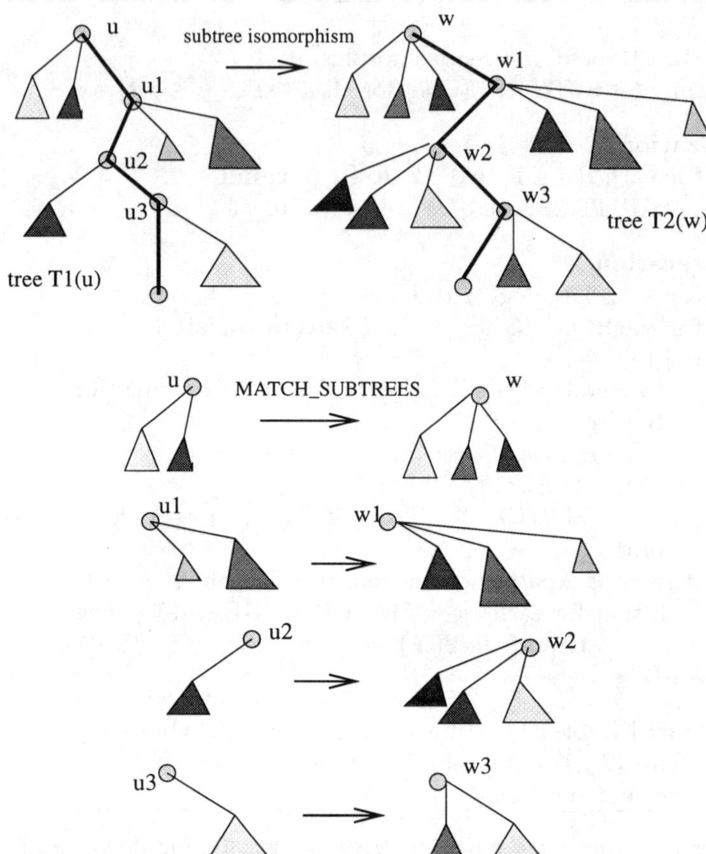

Fig. 14.3 Reduction of the rooted subtree isomorphism problem SUBTREE(u, w) to several independent instances of MATCH_SUBTREES. The heaviest paths from u and w are removed (the nodes remain), then we have several instances of MATCH_SUBTREES. The hanging subtrees are small and the table SUBTREE is already computed for all nodes u which are not on the heaviest path.

A network is a directed graph $G = (V, E)$ with nonnegative integer *capacities* assigned to its edges:

$$c\colon E \to Z^{+}.$$

We distinguish two special nodes $s, t \in V$, called the *source* and the *sink* of the network. Assume there are no edges coming into s and no edges going out from t.

A *flow* is a function assigning nonnegative integers to the edges:

$$f\colon E \to Z,$$

satisfying the conditions:

(1) the flow in each edge does not exceed the capacity of the edge;
(2) the sum of the flows of incoming edges is equal to the sum of the flows of outgoing edges for each node $v \in V - \{s, t\}$.

The value of the flow (denoted by flow_value(f)) is the sum of flows of incoming edges of t, which is equal to the sum of outgoing edges of s. The *network flow* problem consists in computing a flow f maximizing flow_value(f).

Define the relation $\rightsquigarrow$ for edges (u, v), $(u', v') \in E$ as follows:

$$(u, v) \rightsquigarrow (u', v') \Leftrightarrow v = u'.$$

We say that a network is a *unit network* if the capacities of all edges are 1.

For a unit network $G = (V, E)$ we construct a bipartite graph $G' = (X, Y, E')$, where X, Y are disjoint copies of the set of edges of G. The construction is demonstrated in Figure 14.4.

Formally

- $X = \{\, copy_1(e) \colon e \in E, e$ is not of the form $(u, t)\,\}$,
  $Y = \{\, copy_2(e) \colon e \in E, e$ is not of the form $(s, u)\,\}$;
- $(copy_1(e), copy_2(e')) \in E' \iff (e \rightsquigarrow e')$ or $((e = e')$ and e is not incident to s nor $t)$.

The edges of the type $(copy_1(e), copy_2(e))$ are called *special edges*. We assign weight 0 to the special edges of G' and weight 1 to all other edges.

Theorem 14.2.2

The maximum flow problem in networks with polynomially bounded capacities is in RNC. It is in NC if the minimum perfect matching problem for unary-weighted bipartite graphs is in NC.

Proof Assume at the beginning that the network G is a unit network. Let M be a maximum cardinality minimum weight matching in G'. Then for each edge e of G we set $f(e) = 1$ iff e is an endpoint of a matched nonspecial edge in G'. Otherwise we set $f(e) = 0$. The obtained flow has maximum value. The construction is demonstrated in Figure 14.4.

If the network is not a unit network then each edge $e = (u, w)$ of capacity $c(e)$ is replaced by $c(e)$ disjoint unit-capacity edges connecting u and w. We can place on each of these edges an additional *middle* node, to prevent the graph from becoming a multigraph. If the capacities are polynomial then we reduced the flow problem to a problem for a unit network. Then we apply the algorithm for unit networks.

□

Remark Inclusion of the maximum flow problem with general binary weights in the class *RNC* is very unlikely, unless $P = RNC$. However, the following partial result is known.

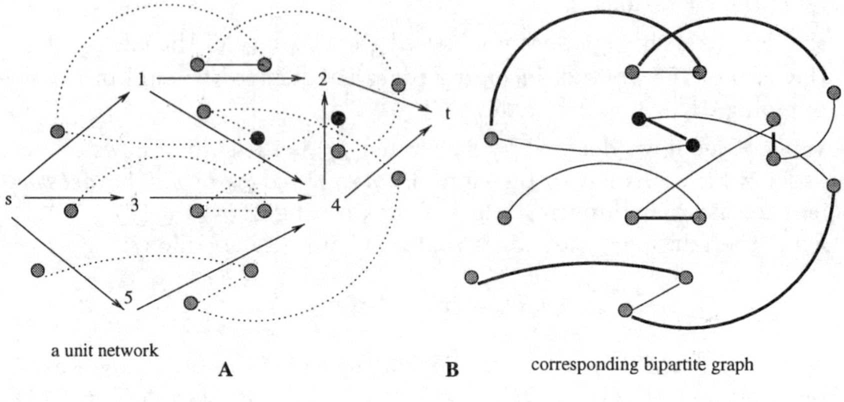

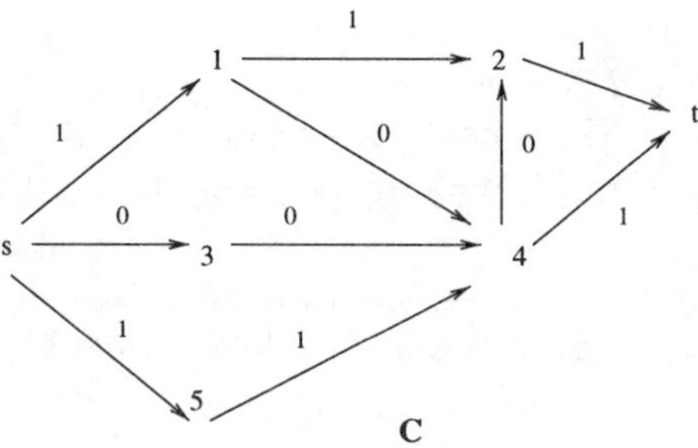

Fig. 14.4 A: a unit network; additional shaded nodes (placed at the edges) correspond to "copies" of edges (only one copy for edges incident to s or t) and the dotted lines correspond to the graph G'. B: the graph G' and a maximum cardinality minimum weight matching (bold edges). C: a maximum flow corresponding to the matching.

Theorem 14.2.3 ([Sp93])
For each constant k there is an RNC-algorithm constructing a maximum flow in networks with arbitrary weights with approximation factor $1 + 1/k$.

Assume G is an undirected *weighted graph* and each edge has an integer nonnegative weight which is polynomially bounded. For two disjoint sets of nodes of an undirected graph G denote by Max_Min_Paths_Between(L, S) a maximum cardinality set of disjoint paths between L and S of minimum total weight. The algorithm for the disjoint paths problem will be the main routine in the next section; however, we need a somewhat weaker version. Assume the following *restriction* whenever dealing with Max_Min_Paths_Between problem in this chapter:

> only the edges incident to L can have nonzero weights, all the others have zero weight.

The problem Max_Min_Paths_Between can be reduced to a network flow problem with polynomially bounded capacities, see [E79]. This fact and Theorem 14.2.2 imply the following theorem.

Theorem 14.2.4
The Max_Min_Paths_Between *problem in graphs which have polynomially bounded weights is in RNC. It is in NC if the unary-weighted minimum perfect matching problem for bipartite graphs is in NC.*

Aggarwal and Anderson have given in [AA88] an interesting algorithm avoiding the network flow approach for the Max_Min_Paths problem by reducing it directly to a minimum weight perfect matching. We briefly describe their construction below. In fact it is very similar to the construction for maximum flows.

We start with the unweighted case. Denote by Max_Paths_Between the problem of computing a maximum cardinality set of disjoint unweighted paths between two disjoint sets L and S of nodes of an undirected graph G.

Lemma 14.2.5
The problem Max_Paths_Between *in undirected graphs is NC-reducible to a minimum perfect matching problem in weighted graphs with zero-one weights.*

Proof Construct a new graph $G' = (V', E')$. Assume $|L| = |S|$; if not then we can add some *dummy* nodes. Assume also that L and S are sets of independent nodes (no edge between two nodes which are both in L or both in S).

Define $\hat{V} = V - (L \cup S)$. Each node v (except the ones in $L \cup S$) is split into two copies. Formally we define G' as follows:

$$\begin{aligned}
V' &= L \cup S \cup \{v_{\text{in}} : v \in \hat{V}\} \cup \{v_{\text{out}} : v \in \hat{V}\} \\
E'' &= \{(v_{\text{in}}, w_{\text{out}}) : (v, w) \in E,\ v, w \in \hat{V}\} \\
&\quad \cup \{(x, y) : (x, y) \in E,\ x, y \in (L \cup S)\} \\
&\quad \cup \{(v_{\text{in}}, v_{\text{out}}) : v \in \hat{V}\} \\
&\quad \cup \{(x, v_{\text{in}}) : x \in L, v \in \hat{V},\ (x, v) \in E\} \\
&\quad \cup \{(v_{\text{out}}, y) : y \in S, v \in \hat{V},\ (v, y) \in E\}.
\end{aligned}$$

The set E' of edges of G' consists of edges in E'', whose weight is 0, and of additional edges (x, y) such that $(x, y) \notin E$ and $x, y \in (L \cup S)$. We call them *additional* edges and assign weight 1 to them.

We describe how a minimum perfect matching in G' corresponds to a maximum set of disjoint paths in G.

Claim Suppose we have a perfect matching M of weight $|L| - j$. Denote $M' = \{(v_{\text{in}}, v_{\text{out}}) : v \in \hat{V}\}$. Let $\oplus$ denote symmetric difference. Then the set of paths in $M \oplus M'$ of weight 0 is a maximum set of disjoint paths between L and S.

Proof (of the claim). Consider the graph G'' formed by edges in $M \oplus M'$. It consists of paths and cycles. Each node in $\hat{V}$ has degree 0 or 2 in G''. Hence the paths in $M \oplus M'$ correspond to disjoint paths between L and S in G. On the other hand if we have a set of j disjoint paths in G then we can construct in Gi a perfect matching of weight $|L| - j$.

Choose corresponding paths in G' and take each second edge from each such path. Observe that we can start with an edge covering a node in L and end with an edge covering a node in S, using the trick of doubling nodes and putting edges in the form $(v_{\text{in}}, v_{\text{out}})$. For all nodes of L, S not covered in this way we can take additional edges ($|L| - j$ such edges is enough).

Hence the set of zero-weight (open) paths in G'' gives a maximum set of disjoint paths. This completes the proof of the claim.

The thesis of the lemma follows directly from the claim.

□

We now show how to solve the problem Max_Min_Paths_Between using the algorithm for Max_Paths_Between. Assume we know that the maximum cardinality of a set of disjoint paths between L and S equals j. Now we can find a minimum weight set of j disjoint paths.

Lemma 14.2.6
Assume j is given. The problem of finding a minimum weight set of j disjoint paths between L and S can be reduced to a minimum weight perfect matching.

Proof We do the same construction as above, but instead of adding *additional* edges between L and S we proceed as follows.

Create two sets L' and S' of cardinality $|L| - j$. We add all possible edges between L and L' and between S and S' creating two complete bipartite subgraphs with all zero-weigh-edges. The for each edge of the form (x, v_in), where $x \in L$, we give it the same original weight as in G. Observe that only edges incident to L in G can have nonzero weights (see the definition of the Max_Min_Paths_Between problem and the restriction which we have assumed).

Now the minimum weight perfect matching for a modified graph G' gives a set of j disjoint paths of minimum cost in the same way as in Lemma 14.2.5.

□

14.3 Parallel construction of DFS-trees

A *DFS-tree* is one of the main data structures in graph theory. Assume G is an undirected connected graph with n vertices. A *DFS*-tree is a rooted directed tree T containing all nodes of G and having the following *DFS-property*: each edge of G is between two nodes lying on the same top-down branch of T; in other words if (u, w) is an edge then u is an ancestor of w or w is an ancestor of u in T.

The edges violating the DFS-property are called *crossing edges* for T, so T is a DFS-tree if there are no crossing edges w.r.t. T.

We can similarly define a *crossing path* with respect to a tree T as a path consisting of edges which are not in T and which connect two nodes of T, none of which is an ancestor of the other.

We say that

> a set S is a *good separator* of an n-vertex graph G iff the largest connected component of $G - S$ has size at most $\frac{n}{2}$.

The main function in the DFS-construction is the Good_Path_Separator(G) function, which returns a good separator P consisting of at most 11 disjoint paths.

We say that

> a tree T is a partial DFS-tree iff T is a directed rooted tree not necessarily containing all nodes of G and there is no crossing path w.r.t. T.

The auxiliary function is Partial_DFS_Tree (P, v). If P is a good path separator of cardinality at most 11 then the function returns a partial DFS-tree T' rooted at v, all nodes of paths in P are contained in T', but it is possible that T' contains some other nodes as well.

```
function Partial_DFS_Tree(P, v);
{returns a partial DFS-tree T' containing all nodes
of P and rooted at the node v;
assume P is a set of at most 11 disjoint paths}
initially T' is a tree consisting of a single node v;
while P contains some nonempty path π do
    begin
    {invariant: there are no crossing edges w.r.t. T'}
    find a lowest node v ∈ T' from which there
    is a path γ to some node w ∈ π;
    decompose path π as π = π1wπ2, where |π1| ≥ |π2|;
    add the path π1γ to the tree T';
    π := π2  {possibly π = ∅};
    end
return T';
```

The algorithm for the full DFS-tree has a *divide-and-conquer* structure. First a good and small path separator P is computed. Then, in parallel, the Partial_DFS_Tree(P) and DFS-trees are computed for each connected component of $G - P$. The next step is to merge all computed trees together.

Algorithm Parallel_DFS_Tree(G, v);

$P :=$ Good_Path_Separator(G, v);
$T' :=$ Partial_DFS_Tree(P, v);
compute connected components $C_1, C_2, \ldots, C_k$ of $G - P$;

for each component C_i **do in parallel**
  begin
  let (v_i, u_i) be an edge between C_i and T'
  such that u_i is the lowest possible node of T';
  $T_i :=$ Parallel_DFS_Tree(C_i, v_i);
  connect T' to T_i by the edge (u_i, v_i);
  end

return T';

The structure of the algorithm is illustrated schematically in Figure 14.5.

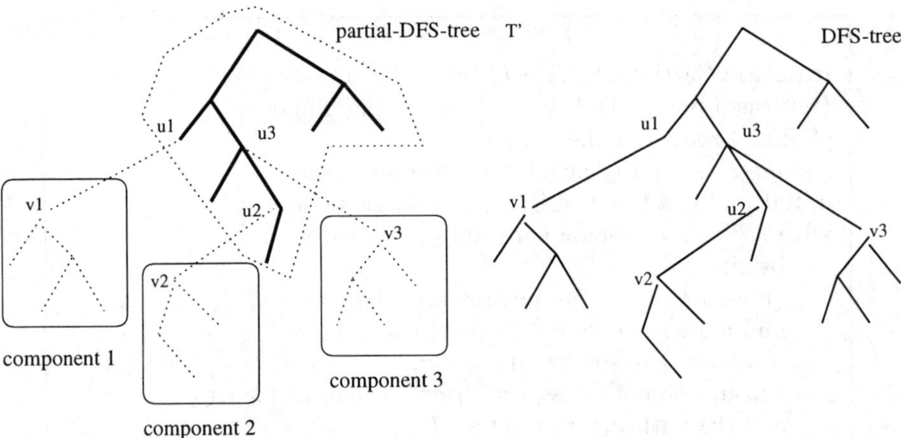

Fig. 14.5 A full DFS-tree is constructed by merging DFS-trees of connected components (computed independently) with the partial DFS-tree T'.

We now pass to the crucial part of the DFS-tree construction: implementation of the function Good_Path_Separator(G).

At the heart of the small separator construction is the procedure of *separator reduction* called REDUCE(P).

The value of the parameter P in REDUCE is always a good path separator but it is not necessarily very small. Denote by $k = |P|$ the number of paths in P, so P consists of disjoint paths and the largest component of $G - P$ has size bounded by $\frac{n}{2}$.

The initial good separator P_0 is rather trivial; it consists of all single nodes of G treated as one-node paths.

The procedure REDUCE(P) decreases $|P|$ by a factor of at least $\frac{11}{12}$, if $|P| > 11$. Eventually after a logarithmic number of applications of REDUCE we obtain a good path separator which is of at most size 11.

```
function Good_Path_Separator(G);

P := the set of all single nodes of G treated as one-node paths;
while |P| > 11 do
        P := REDUCE(P);

return P;
```

The set of paths P is partitioned into two disjoint sets L and S. The letter S stands for "short paths", since the paths in S will be shortened in some stages. Eventually the set S disappears or becomes very small.

Initially

$$|L| = \frac{1}{4}|P| \text{ and } |S| = \frac{3}{4}|P|.$$

Consider two paths $\alpha \in L$ and $\beta \in S$ and a path $\gamma \in G - P$ connecting a point $x \in \alpha$ and a point $y \in \beta$, see Figure 14.6. Assume the structure of paths is as follows:

$$\alpha = \alpha' x \alpha'' \text{ and } \beta = \beta' y \beta'',$$

where $|\alpha'| \geq |\alpha''|$ and $|\beta'| \geq |\beta''|$.

Denote $cost(\gamma) = |\alpha''|$.

The segment α'' is to be potentially discarded, so minimizing the length of the discarded segment corresponds to minimizing the cost of the path γ.

Define the following sets:

$$\begin{aligned} \text{newpath}(\gamma) &= \alpha'\gamma\beta', \\ \text{left\_path}(\gamma) &= \alpha'', \\ \text{right\_path}(\gamma) &= \beta''. \end{aligned}$$

Let $R = \text{Max\_Min\_Paths\_Between}(L, S)$ be a maximum cardinality set of disjoint paths in $G - P$ between paths in L and paths in S which is a minimal total cost (in the sense defined above) among sets of maximum cardinality. Define several sets depending on L, S and R to simplify the notation the dependence on R will not be included explicitly into the names of the defined sets. The sets of paths are illustrated in Figure 14.7.

- $\mathit{NewPaths} = \{\text{newpath}(\gamma) : \gamma \in R\}$;

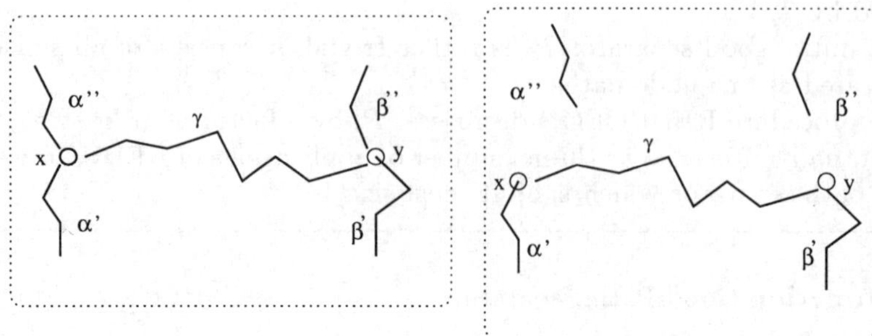

Fig. 14.6 Two paths $\alpha \in L$ and $\beta \in S$ are connected by a path $\gamma \in G-P$. On the right we show how the three new paths are created: left_path(γ) $= \alpha''$, right_path(γ) $= \beta''$ and newpath(γ) $= \alpha'\gamma\beta'$. The cost of γ is the length of α'' which is to be minimized.

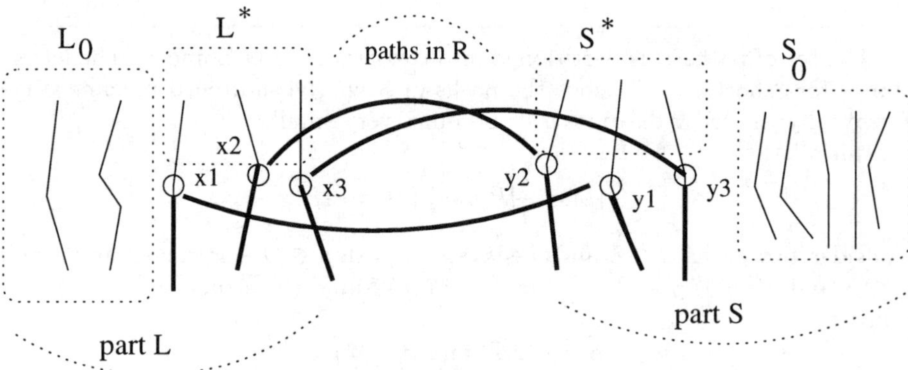

Fig. 14.7 The graphical illustration of sets of paths. R consists of paths $x1 \leftrightarrow y1$, $x2 \leftrightarrow y2$, $x3 \leftrightarrow y3$. The set *Newpaths* consists of paths indicated in bold.

- $L^* = \{\text{left\_path}(\gamma) : \gamma \in R\}$;
- $S^* = \{\text{right\_path}(\gamma) : \gamma \in R\}$;
- L_0 and S_0 are subsets of all paths in L, S, respectively, which are not connected by any path in R

In one parallel operation we merge together some paths in L, S and R and obtain the new set of paths *Newpaths*. However, it may be difficult to disregard other paths in L and S, since the property of a good separator could be violated. We also need to discard something to reduce the size.

Fortunately there is always the possibility to choose at least one good combination out of the following four:

1. $New_1 = L_0 \cup Newpaths \cup S^* \cup S_0$.
2. $New_2 = L_0 \cup Newpaths \cup L^* \cup S^*$.
3. $New_3 = Newpaths \cup L^* \cup S^* \cup L_0$.

4. $New_4 = Newpaths \cup L^* \cup S^* \cup S_0$.

Lemma 14.3.1 (key lemma)

1. *New_1 or New_2 is a good separator.*
2. *If $|R| < \frac{1}{12}k$ then New_3 or New_4 is a good separator.*

Proof Both points are proved using the following simple claim.

Claim A

If the largest connected component of X is not larger than $n/2$ and there is no path in X between Y and Z then

- the largest connected component of $X \cup Y$ does not exceed $n/2$, or
- the largest connected component of $X \cup Z$ does not exceed $n/2$.

Using the claim we prove each point separately.

1. We start the proof of point 1 with the following observation which follows from the maximum cardinality and minimum weight of R.
   The crucial property of new sets of paths is

   there is no path in $G - (P \cup R)$ between L^* and S_0.

   Apply Claim A now with $X = G - (P \cup R)$, $Y = L^*$ and $Z = S_0$. Observe that $G - New_1 = X \cup Y$ and $G - New_2 = X \cup Z$.
   Consequently New_1 or New_2 is a good separator (there is no large component in the remaining part of the graph). This completes the proof of point 1.
2. Assume now that $|R| < |P|/12$. Then there is no path in $G-(Newpaths \cup L^* \cup S^*)$ between L_0 and S_0. Applying Claim A we obtain in this case that New_3 or New_4 is a good separator.

This completes the proof of the key lemma in this section.

□

We need to implement the function REDUCE to reduce the size of P by the factor $\frac{11}{12}$. The lemma suggests that we can set P to one of New_i. It is easy to calculate that for $i \in \{4, 2, 3\}$ we get the required reduction.

However, if New_1 is a good separator then the size of P does not change. In this case a large proportion of paths in S is reduced by at least $\frac{1}{2}$. This implies that after a logarithmic number of applications of $P := New_1$ sufficiently many paths in S disappear (become empty) and we eventually get the reduction $\frac{11}{12}$.

We leave the technical details of such calculations to the reader. The implementation of REDUCE can be summarized as follows.

```
function REDUCE(P);
{|P| > 11}

k := |P|; partition P into sets L and S;
while |P| > (11/12)k do
      begin
          R := Max_Min_Paths_Between(L, S);
          compute New_i for 1 ≤ i ≤ 4;
          if |R| ≥ (1/12)|P| then P := New_i, where i ∈ {1, 2}
          else P := New_i, where i ∈ {3, 4}
          {each time we choose an i to guarantee that P
          is a good separator, this is possible
          due to Lemma 14.3.1};
      end;

return P;
```

Recall that Max_Min_Paths_Between(L, S) is a maximum cardinality set of disjoint paths in $G - P$ between paths in L and paths in S which is of a minimal total cost. Its computation is the crucial point in the function REDUCE and the whole algorithm. All other parts can be done by NC-computations. We have seen in Section 14.1 that Max_Min_Paths_Between(L, S) can be reduced to the computation of a maximum cardinality, minimal weight set of disjoint paths.

We have seen in Section 14.2 that this is reducible to a maximum cardinality matching; therefore it is in RNC. Consequently we have the main result of this section.

Theorem 14.3.2
The construction of a DFS-tree is in RNC. It is in NC if the minimum weight unary-weighted perfect matching problem is in NC.

14.4 Bibliographic notes

The NC-reducibility between subtree isomorphism and perfect matching was shown in Karpinski and Lingas [KL89]. The maximum flow problem for small capacities of edges and its relation to matching was investigated in Karp, Upfal and Wigderson [KUW86]. For the DFS-tree construction algorithm see Aggarwal and Anderson [AA88].

15

More applications

We now sketch more applications of parallel algorithms for matching problems. We show that owing to the existence of *RNC*-algorithms for maximum matchings there exist *RNC*-algorithms for the following problems: bandwidth approximation, superstring approximation, cycle cover, approximate Metric TSP, unary Chinese postman problem, two-processor scheduling and the approximate Steiner tree problem.

15.1 Unary Chinese postman problem

Consider the following problem. A postman is supposed to deliver the mail with in his part of the city. He starts at a post office and has to walk along each street at least once. Finally he has to return to the post office. We model the street graph as an undirected graph. For simplicity assume that each street has the same length (the *unary* version). Assume the graph is connected. We want to find the postman's route which minimizes the length of the route. Observe that some streets have to be traversed more than once, unless the graph is Eulerian. We call this problem the *unary postman problem*.

The problem is equivalent to the following one:

> add a minimal number of edges to a connected undirected graph G to make it a Eulerian multigraph: the degree of each node becomes even.

There is a very simple algorithm:

1. Create a weighted undirected graph G'. The nodes of G' are all odd-degree nodes of G. There is an edge between nodes $u, v \in G'$ iff there is a path between u and v in G (so G' is a complete graph, since G is connected).
2. Set the weight of each edge $(u, v) \in G'$ to the shortest distance from u to v in G.
3. Find a minimum cost perfect matching M in G'.
4. For each edge $(u, v) \in M$ add to G *additional* edges along a shortest path from from u to v; this can cause some multiple edges to be created.

The construction is demonstrated in Figure 15.1.

Theorem 15.1.1
The Chinese postman problem with unary weights is in RNC.

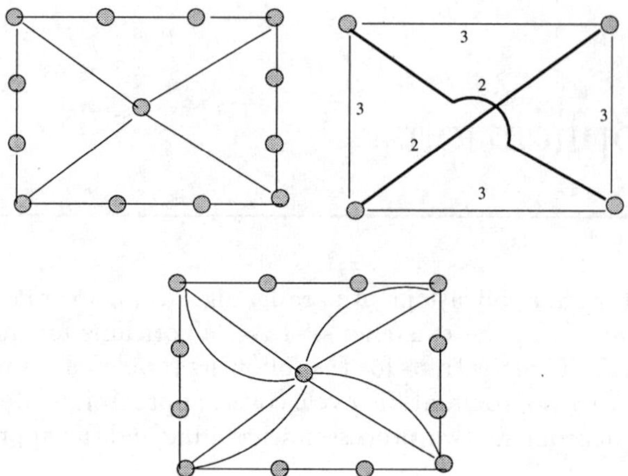

Fig. 15.1 A graph G of streets, the distance graph G' with minimum perfect matching corresponding to bold lines and a solution to the unary postman problem for G.

Proof The computation of distances and shortest paths can be easily done in *NC*. Hence the problem is in *RNC*, by the application of the minimum weight perfect matching problem with small weights.

□

15.2 Two-processor scheduling

An instance of the two-processor scheduling problem is a directed *transitively closed* acyclic graph D, called the *precedence graph* . The transitive closure means that there is a directed edge between two nodes whenever there is a directed path between them. The nodes of D correspond to certain unit-time tasks which are to be computed, and the edge (u, v) of the graph corresponds to the relation "the task u should be computed before v".

A *schedule* is a mapping

$$time\colon \not\leftrightarrow [1 \ldots \text{max\_time}]$$

of tasks (nodes) to nonnegative integers. The mapping is correct iff the following conditions hold:

1. $|time^{-1}(t)| \leq 2$ for each $1 \leq t \leq \text{max\_time}$;
2. $(u, v) \in D$ implies $time(u) < time(t)$.

The *cost* of the schedule D equals max_time.

Lemma 15.2.1 ([FKN69])
The minimum cost of the schedule $D = (V, E)$ equals $|V| - |M|$, where M is a maximum matching in the complement graph $\bar{D}$ of an undirected version of D.

Figure 15.2 shows the direct correspondence of an optimal schedule to a matching in $\bar{D}$. However, not every maximum cardinality matching in $\bar{D}$ *directly*

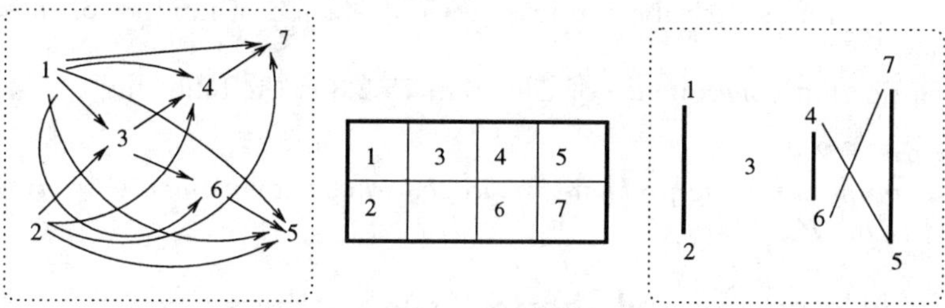

Fig. 15.2 A schedule graph D, an optimal schedule of tasks and the corresponding matching in $\bar{D}$.

gives an optimal schedule for D, but it was proved in [FKN69] that it can be suitably transformed.

Theorem 15.2.2
The computation of the minimum cost of an instance of the two-processor scheduling problem is in RNC.

Proof The problem is NC-reducible to a maximum cardinality matching by Lemma 15.2.1.

□

Helmbold and Mayr have given a stronger result in [HM87], but the proof is quite complicated. In fact the computation of the cost of the schedule is not too complex, and the proof shows a striking difference between the complexity of the decision (in the sense of computing the cost) and construction (in the sense of constructing a minimum cost solution) versions of the same problem.

There are two drawbacks with the algorithm: it is too complex and the number of processors is $O(n^{10})$, which is too large. Hence RNC-algorithms are much more efficient in this case.

Theorem 15.2.3 ([HM87])
The construction of a minimum cost two-processor schedule is in NC.

Proof The problem is NC-reducible to a maximum cardinality matching due to Lemma 15.2.1.

□

We cite also here the following more *philosophical* statement from [HM87]:

> our results on two-processor scheduling provide evidence that the matching problem might be in *NC*, since the two problems are closely related.

We say that an undirected graph is *transitively oriented* iff its edges can be oriented to produce an acyclic transitively oriented graph. A typical family of transitively oriented graphs form the class of complements of interval graphs.

The interval graphs describe the intersection relation of families of intervals on the same line.

An important consequence of Theorem 15.2.3 is the following.

Theorem 15.2.4
The maximum matching problem for graphs whose complements are transitively oriented is in NC.

15.3 Cycle covers and shortest superstrings

A cycle cover is a set of disjoint cycles containing all nodes of a given directed graph. Assume that the graph is weighted (each edge has a nonnegative weight assigned to it). A minimum cycle cover is the one minimizing the total sum of weights of its edges.

Theorem 15.3.1
The minimum cycle cover problem for a digraph with edge weights polynomially bounded is in RNC. If the minimum weight perfect matching problem is in NC for graphs with weights bounded polynomially then the minimum cycle cover with polynomially bounded weights is also in NC.

Proof For a directed graph $G = (V, E)$ we construct a bipartite graph $G' = (X, Y, E')$, where X, Y are disjoint copies of the set of nodes of G.

Formally

- $X = \{copy_1(v) : v \in V\}$, $Y = \{copy_2(v) : v \in V\}$;
- $(copy_1(v), copy_2(w)) \in E' \Leftrightarrow (v, w) \in E$.

Each perfect matching of G' gives a set $\bar{E} \subseteq E$ such that the indegree and outdegree of each node with respect to $\bar{E}$ equals one. In other words $\bar{E}$ is a cycle cover of G. A minimum weight cycle cover therefore corresponds to a minimum weight perfect matching.

□

Let $S = \{s_1, \ldots, s_n\}$ be a set of n strings over some alphabet Σ. A *superstring* of S is a string sp over Σ such that each string $s_i \in S$ appears as a substring of sp.

The *shortest superstring* problem is to find for a given set S the shortest superstring ss(S) . We use opt(S) to denote the length of ss(S) . Assume further, without loss of generality, that no string $s_i \in S$ is a substring of any other $s_j \in S$.

It is known that the shortest superstring problem is $\mathcal{NP}$-hard. Because of its important applications in data compression and DNA sequencing, it is of interest to find approximation algorithms with good performance guarantees, see [CR94].

To evaluate the effectiveness of the obtained approximation the following measure is used. The *superstring approximation* problem is to find a superstring sp of S such that the ratio $|sp|/\operatorname{opt}(S)$ is minimized. We will call this ratio the *approximation factor* of a superstring.

For two strings s and t let v be the longest string such that $s = uv$ and $t = vw$ for some nonempty strings u and w. Denote $\operatorname{dist}(s, t) = |u|$. For a set S

of strings let G_S be the directed graphs whose nodes are elements (strings) of S and the weight of an edge (s,t) is $\text{dist}(s,t)$. For a path $\pi = (s_1, s_2, \ldots, s_k)$ in the graph G_S denote

$$\text{string}(\pi) = u_1 u_2 u_3 \ldots u_{k-1} s_k$$

where u_i is the prefix of length $\text{dist}(s_i, s_{i+1})$ of s_i.

Observation If π is the shortest Hamiltonian path in G_S then $\text{string}(\pi)$ is the shortest superstring.

Unfortunately the Hamiltonian path problem is *NP*-complete, but it happens that in a certain sense we can replace the Hamiltonian path by a minimum cycle cover.

The following simple algorithm was introduced in Blum, Jiang, Li, Tromp and Yannakakis [BJLTY91].

Algorithm Concat-Cycles (S);
construct the graph G_S;
compute a minimum weight cycle cover
$\quad\quad \mathcal{C} = \{C_1, C_2, \ldots, C_r \text{ in } G_S\}$;
for each cycle $C_i \in \mathcal{C}$ **do in parallel**
$\quad\quad \pi_i := C_i$ with an arbitrary edge removed;
$\quad\quad w_i := \text{string}(\pi_i)$;
return $w_1 w_2 \ldots w_r$;

Theorem 15.3.2
There is an RNC-algorithm which constructs the shortest superstring with approximation factor 4.

Proof The algorithm Concat-Cycles is in *RNC* owing to Theorem 15.3.1 and to the fact that weights are small: $\text{dist}(s,t) \leq n$. It was shown (by a complicated argument) in [BJLTY91] that the approximation factor of the algorithm is 4. □

The approximation factor was later improved to 2.83 in [CGPR], and then to 2.793 in [C95]. The computations depend on *RNC*-algorithms for matchings and are based on the same idea as the algorithm Concat-Cycles.

There is also a nice application of *maximal* matchings to superstrings. The *greedy* sequential algorithm is parallelized using maximal sets of paths.

The set P is a *maximal path set* (MPS, in short) in a graph iff it consists of several disjoint paths and is inclusion maximal among sets with this property. Chen has shown in [C95] how to compute an MPS by an *NC*-algorithm. The algorithm makes a logarithmic number of iterations, and in each one the smaller sets are merged in parallel into larger ones. This is done by constructing a graph of possible merges and computing a maximal matching in such a graph. This matching corresponds to path merges which can be done independently. We refer the reader to [C95] for details of this algorithm. It was used to approximate superstrings with respect to a *compression factor* defined as the ratio $(|S| - |S|')/(|S| - \text{opt}(S))$, where S' is the superstring produced by our algorithm.

Theorem 15.3.3 ([CGPR], [C95])
There is an NC-algorithm which constructs a superstring with compression ratio $1/(3+\epsilon)$.

15.4 The $\frac{3}{2}$-approximation of the metric TSP

The *metric travelling salesman problem* (Δ-TSP, in short) is the problem of finding the *shortest* Hamiltonian cycle in a weighted undirected complete graph G satisfying the *triangle inequality*

$$\text{dist}(u, v) \leq \text{dist}(u, w) + \text{dist}(w, v),$$

for each $u, v, w \in G$.

The triangle inequality allows us to make so-called *short-cuts.* Assume we have an Euler cycle C' and go around this cycle. If we go from v to w along some path of previously visited nodes then we can go directly through one edge (u, w) from u to w. By the triangle inequality the length of a single edge (u, w) is not larger than the length of any path from u to w.

There is a simple $\frac{3}{2}$-approximation algorithm for the Δ-TSP given by Christofides in [C76]. The algorithm is similar to the one for the postman problem and is easily parallelized.

Theorem 15.4.1
The $\frac{3}{2}$-approximation of the metric TSP with distances bounded polynomially is in RNC.

Proof The Christofides algorithm works in several stages:

1. Find a minimum spanning tree T of G by an NC-algorithm, see [GR88].
2. $X :=$ set of nodes of T of odd degree.
3. Compute a minimum weight perfect matching M in the graph induced by X.
4. $G' := T \cup M$.
5. Find an Euler tour Euler_Cycle of G', see [GR88].
6. Return the Hamiltonian cycle C of G by traversing Euler_Cycle and making short-cuts.

All stages in the algorithm can be done in NC except matching. If the distances are polynomially bounded then stage 3 can be done by an RNC-algorithm.

□

In a major breakthrough, recently Arora [A96] improved the above significantly for the case of a Euclidean norm in showing the existence of polynomial time approximation algorithms with arbitrary small constant approximation ratios (PTAS) for this case.

15.5 The bandwidth problem

Let $G = (V, E)$ be a simple graph on n vertices. A layout of G is a one-to-one mapping $f: V \to \{1, \ldots, n\}$. The bandwidth $b(f, G)$ of this mapping is defined by

$$B(f, G) := \max\{|f(v) - f(w)|: \{v, w\} \in E\},$$

the greatest distance between adjacent vertices in G corresponding to f. The bandwidth $B(G)$ is then

$$B(G) := \min_{f \text{ is layout of } G} \{B(f, G)\}.$$

Clearly the bandwidth of G is the greatest bandwidth of its components.

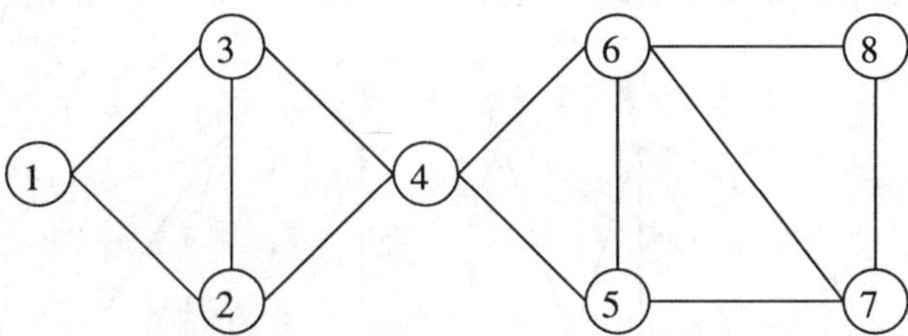

Fig. 15.3 A small 1/4-dense graph G. It has eight vertices and minimum degree 2.

The problem of finding the bandwidth of a graph is NP-hard even for trees with maximum degree 3 [GGJK78]. It is not much known about its approximation hardness (cf. also recent results of Karpinski and Wirtgen [KW97b], Blache , Karpinski and Wirtgen [BKW97], and Feige [F97]). Recently Karpinski , Wirtgen and Zelikovsky [KWZ97] found a 3-approximation RNC-algorithm for the bandwidth problem restricted to ϵ-dense graphs. We call a graph G ϵ-dense (cf. also Arora , Karger and Karpinski [AKK95]), Karpinski [K97]) if the minimum degree $\delta(G)$ is at least ϵn. Here we sketch a weaker version of this algorithm, namely a 4-approximation RNC-algorithm. It uses as one building block the construction of a perfect matching in a bipartite graph.

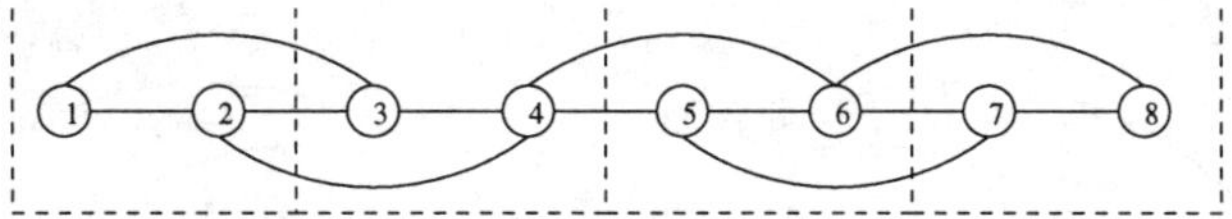

Fig. 15.4 An optimum layout of the graph G in Figure 15.3. It is optimum because $\delta(G) = 2$ and the maximum distance of two neighboring vertices is 2.

If we have some optimum layout, we can split this layout into $n/B(G)$ boxes, so that there are only edges between neighboring boxes (see Figur 15.4). It is clear that a graph with minimum degree k has at least bandwidth k. Therefore the bandwith of ϵ-dense graphs is at least ϵn and thus we have at most $1/\epsilon \in O(1)$ boxes.

The algorithm chooses at random $O(\log n)$ vertices $S \subseteq V$, the so-called special vertices. We have two important properties of S:

1. S forms a dominating set with high probability,
2. There is a high probability that each of the boxes has at least one representative in S.

Now suppose we know to which box each special vertex belongs. (In fact we can find the right assignment of the special vertices to the boxes by an exhaustive search with polynomial bounded work.) For any vertex which is not special we have at most three possible boxes where it might belong (see Figure 15.5).

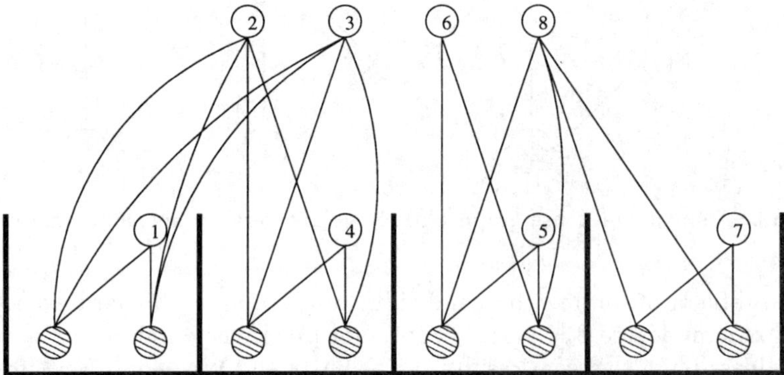

Fig. 15.5 The vertices $\{1, 4, 5, 7\}$ are randomly chosen special vertices. In the figure we see the right assignment of these vertices to the boxes. For all the neighbors of the special vertices we know the area of at most three boxes to which they belong.

As in Figure 15.5 we construct an auxiliary graph in which each vertex of the input graph is conected to the possible places in the boxes. Clearly a perfect matching in this graph gives us a layout (see Figure 15.6).

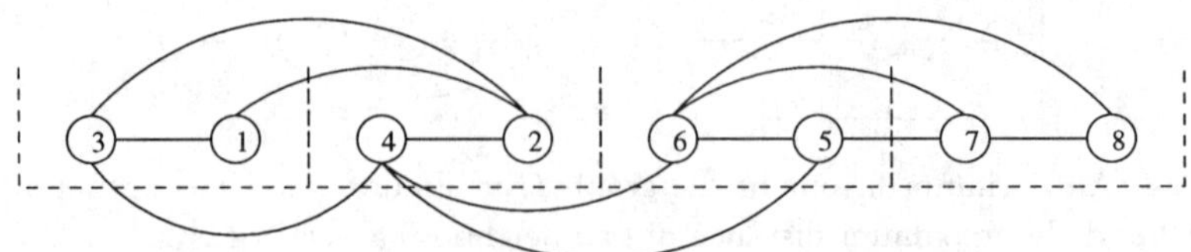

Fig. 15.6 After running the perfect matching algorithm, we get a layout with a maximum distance of at most $3 \leq 2B(G)$.

The worst case arises in the following situation: there are two non-special vertices u and v which are adjacent. The dominating special vertices s and t lie in neighboring boxes 2 and 3. The perfect matching algorithm now assigns u to the far left side of box 1 and v to the far right side of box 4 (see Figure 15.7)

We can summarize the above discussion in the following algorithm.

```
Algorithm DENSE_BANDWIDTH(G)
{G is ε-dense}
for boxsize = εn to n/2 do
    begin
    {We have ⌈1/boxsize⌉ boxes, being parts of a layout}
    choose at random and independently a subset S ⊆ V of size O(log n);
    for any possible assignment of the vertices of S to the boxes do
        begin
        {build a bipartite auxilary graph G_A of which
         one color class consists of the places in the boxes and
         the other class of the vertices of G}
        for each vertex v ∈ V do
            begin
            if v is in S, connect v to all possible places in its box,
            else it is adjacent to some special vertices s_1, ..., s_k ∈ S;
            build the intersection B_v of the two neighboring boxes
            around each special vertex s_i, i = 1, ..k, including its own box;
            if B_v is empty, we assign the vertices in S to the wrong boxes,
            else connect v to all the places in B_v;
            if there is a perfect matching in G_A, return one of them
            {Note that a perfect matching M also defines a layout f_M}
            end
        end
    end
end DENSE_BANDWIDTH
```

It is clear that we can do all the **for** loops in parallel. At least one of the polynomial number of processes succeeds and gives us a layout which is not too far from the optimum. So we have the following theorem (Karpinski , Wirtgen and Zelikovsky [KWZ97]).

Theorem 15.5.1
There is a RNC-algorithm which finds a layout f for a ϵ-dense graph G, so that $B(f,G) \leq 4B(G)$.

15.6 Spanning 3-hypertrees and approximating Steiner tree

As a last application we mention briefly how the algebraic tools developed in Chapter 5 can be used for another problem in a very similar way to the matching problem.

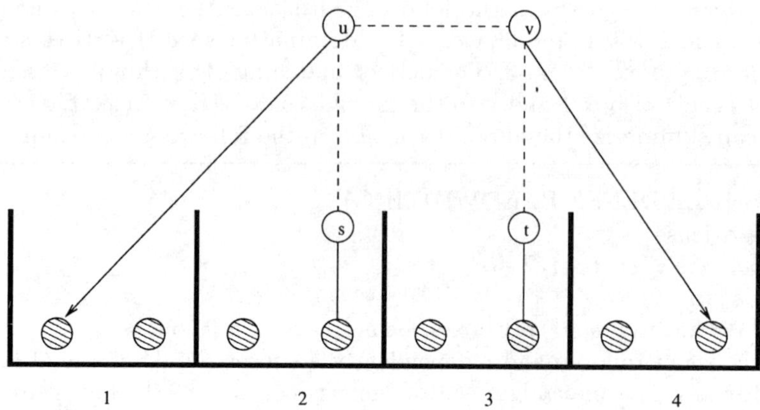

Fig. 15.7 A layout with a bandwidth four times the optimum.

A *hypergraph* $H = (V, F)$ is specified by its node set V and the set F of *hyperedges* which are subsets of V. The hypergraph is k-uniform (it is a k-hypergraph) iff each hyperedge is of size k. Hence the standard graph is a 2-hypergraph.

The spanning tree of an undirected graph is a connected subgraph of G containing all its nodes and no cycles. This definition can be extended naturally to hypergraphs. Two nodes are neighbors in a hypergraph H iff they are members of the same hyperedge. A cycle in a hypergraph is a sequence of distinct neighboring nodes, except the first and the last, which are the same.

A *spanning hypertree* is a subset $F' \subseteq F$ of hyperedges, together containing all nodes of H and no cycles. We say that a hypertree is a k-hypertree iff it is k-uniform. If the hyperedges have weights assigned to them then we define minimum spanning hypertrees as the ones minimizing the total sum of weights of their hyperedges.

The problem of spanning 2-hypertrees (spanning trees for standard graphs) is computationally easy, and there is even a polynomial time algorithm to count exactly the number of all such trees, see [HP73]. This algorithm is yet another demonstration of linear algebra at work: the main tool is again the determinant.

On the other hand the decision version of the same problem for 4-hypertrees is *NP*-complete. The existence version of the spanning 3-hypertree problem is in P, but the deterministic algorithm is very complicated. In [L79a] a randomized algebraic approach was used to produce an RNC-algorithm.

Assume G is a standard graph and there is given an equivalence relation $\equiv$ on edges, each equivalence class being of size 2. Then a spanning tree T of G is called a *double spanning tree* iff $e \equiv e'$ implies $e \in T \Leftrightarrow e' \in T$.

For a 3-hypergraph $H = (V, F)$ we can construct a standard multi graph $G = (V, E)$, such that each spanning hypertree of H corresponds to a double spanning tree of G. For each 3-hyperedge $(v_i, v_j, v_k) \in F$ we choose one node v_k and put the edges (v_i, v_j) and (v_j, v_k) into E.

Then we can construct a *symbolic black-box polynomial* (which we can call a *hypertree symbolic polynomial*) $\mathbf{HP}_G(\bar{x})$. The variables x_{ij} correspond to the edges (v_i, v_j) of G in the same way as in Chapter 6.

Let $\mathcal{T}$ be the family of all double spanning trees of G. $\mathbf{HP}_G$ is a $\mathcal{T}$*-consistent polynomial.* Now we can use our tools from Chapter 5 and prove the following.

Theorem 15.6.1 ([PS97])
We can test for an existence and find (if there is any) a spanning 3-hypertree by an RNC algorithm. If the weights are polynomially bounded then a minimum spanning 3-hypertree can be found by an RNC algorithm.

Theorem 15.6.1 was used in [PS97] to construct an approximately optimal Steiner tree: a subgraph of G spanning a specified subset of nodes. The approximation is close to the best know sequential ratio of Karpinski and Zelikovsky [KZ97].

Theorem 15.6.2 ([PS97])
The problem of the $(1+\epsilon)\frac{5}{3}$*-approximation of Steiner trees is in RNC.*

15.7 Bibliographic notes

The *NC*-algorithm for the algorithm of the two-processor scheduling problem was given in Hembold and Mayr [HM87]. The superstring approximation algorithms are from Czumaj, Gasieniec, Piotrow and Rytter [CGPR], Chen [C95] and Blum, Jiang, Li, Tromp and Yannakakis [BJLTY91]. For the approximation of the bandwidth problem see Karpinski, Wirtgen and Zelikovsky [KWZ97], and for the Steiner tree problem see Prömel and Steger [PS97]. On the approximation hardness of the bandwidth problem see Karpinski and Wirtgen [KW97b], and Blache, Karpinski and Wirtgen [BKW97]. Another interesting applications of parallel matching algorithms can be found in Cheriyan and Thurimella [CT96].

Bibliography

[AA88] A. Aggarwal and R.J. Anderson (1988) A random *NC*-algorithm for depth first search, *Proc. 19th ACM STOC*, 1987, 325–334, also in *Combinatorica*, **8**, 1–12.

[AAK90] A. Aggarwal, R.J. Anderson and M. Kao (1990) Parallel depth first search in general directed graphs, *SIAM J. Comput.*, **19**, 397–409.

[AHU74] A. Aho, J. Hopcroft and J. Ullman (1974) *The design and analysis of computer algorithms*, Addison-Wesley, Reading, Massachusetts.

[ABI86] N. Alon, L. Babai and A. Itai (1986) A fast and simple randomized parallel algorithm for the maximal independent set problem, *J. of Algorithms*, **7**, pp. 567–583.

[ACG93] A. Alon, F. Chung and R. Graham (1993) Routing permutations in graphs via matchings, *STOC'93*.

[A87] R.J. Anderson (1987) A parallel algorithm for the maximal path problem, *Combinatorica*, **7**, 315–326.

[AV79] D. Angluin and L. Valiant (1979) Fast probabilistic algorithms for Hamiltonian circuits and matchings, *J. Comput. Syst.*, **18**, 155–193.

[A96] S. Arora (1996) Polynomial time approximation schemes for Euclidean TSP and other geometric problems, in *FOCS*, pp. 2–11.

[AKK95] S. Arora, D. Karger and M. Karpinski (1995) Polynomial time approximation schemes for dense instances of *NP*-hard problems, *Proc. 27th ACM STOC*, pp. 284–293.

[AIS84] B. Awerbuch, A. Israeli and Y. Shiloach (1984) Finding Euler circuits in logarithmic parallel time, *Proc. 16th ACM STOC*, pp. 249–257.

[BL90] P. Beame and M. Luby (1990) Parallel search for maximal independence given minimal dependence, in *1st ACM Symp. on Discrete Algorithms, SODA*, pp. 212–218.

[B73] C. Berge (1973) *Graphs and Hypergraphs*, North-Holland, New York.

[B84] S. Berkowitz (1984) On computing the determinant in small parallel time using a small number of processors, *Inf. Process. Lett.*, **18**, 147–150.

[BKW97] G. Blache, M. Karpinski and J. Wirtgen (1997) On approximation intractability of the bandwidth problem, *Research Report* No. 85182-CS, University of Bonn.

[B90] N. Blum (1990) A new approach to maximum matching in general graphs, *Research Report* No. 8546-CS, University of Bonn.

[B91] N. Blum (1991) Jack Edmonds original algorithm needs only $O(n^3)$ time, *Research Report* No. 8568-CS, University of Bonn.

[BJLTY91] A. Blum, T. Jiang, M. Li, J. Tromp, and M. Yannakakis (1991)

Linear approximation of shortest superstrings, in *Proc. of the 23rd STOC*, pp. 328–336.

[Bo78] B. Bollobás (1978) *Extremal Graph Theory*, Academic Press, London.

[BGH82] A. Borodin, J. von zur Gathen and J. Hopcroft (1982) Fast parallel matrix and GCD computations, *Proc. 23rd IEEE FOCS*, pp. 65–71.

[Br86] A.Z. Broder (1986) How hard is it to marry at random?, *Proc. 18th ACM STOC*, pp. 50–58.

[CRS93] S. Chari, P. Rohatgi and A. Srinivasan (1993) Randomness-optimal unique element isolation with application to perfect matching, *STOC'93*.

[C95] Z. Chen (1995) *NC*-algorithms for finding a maximal set of paths with application to compressed strings, in *ICALP*, pp. 99–109.

[C94] J. Cheriyan (1994) A Las Vegas $O(n^{2.38})$ algorithm for the cardinality of a maximum matching, in *5th ACM-AMS Symposium on Discrete Algorithms*.

[CT96] J. Cheriyan and R. Thurimella (1996) Approximating minimum-size k-connected spanning subgraphs via matchings, in *FOCS*, pp. 292–301.

[C76] N. Christofides (1976) Worst case analysis of a new heuristic for the travelling salesman problem, *Tech. Report* G.S.I.A., Carnegie-Mellon Univ.

[CNN89] M. Chrobak, J. Naor and M. Novick (1989) Using bounded degree spanning subtrees in algorithms for claw free graphs, in *WADS*, pp. 147–162.

[Ch74] K.L. Chung (1974) *Elementary probability theory*, Springer-Verlag, New York.

[CG72] E.G. Coffman and R.L. Graham (1972) Optimal scheduling for two processors systems, *Acta Inf.*, **1**, 200–213.

[Co74] P.M. Cohn (1974) *Algebra*, vol. 1, John Wiley, New York.

[C85] S. Cook (1985) A taxonomy of problems with fast parallel algorithms, *Inf. Control*, **64**, 2–22.

[CR94] M. Crochemore and W. Rytter (1994) *Text algorithms*, Oxford University Press.

[CGPR] A. Czumaj, L. Gasieniec, M. Piotrow and W. Rytter, Sequential and parallel approximation of shortest superstrings, to appear in *J. Algorithms*.

[DHK93] E. Dahlhaus, P. Hajnal and M. Karpinski (1993) On the parallel complexity of Hamiltonian cycles and matching problem on dense graphs, *J. Algorithms*, **15**, 367–384.

[DK86] E. Dahlhaus and M. Karpinski (1986) The matching problem for strongly chordal graphs is in *NC*, *Research Report* No. 855-CS, University of Bonn.

[DK92] E. Dahlhaus and M. Karpinski (1992) Perfect matching for regular graphs is AC^0-hard for the general matching problem, *J. Comput. Syst. Sci.*, **44**, 94–102.

[DK87] E. Dahlhaus and M. Karpinski (1987) Parallel construction of perfect matchings and Hamiltonian cycles on dense graphs, *Research Report* No. 8518-CS, University of Bonn.

[DK89] E. Dahlhaus and M. Karpinski (1989) A parallel algorithm for maximum matching in planar graphs, *Research Report* No. 8532-CS, University of Bonn.

[DKL89] E. Dahlhaus, M. Karpinski and A. Lingas (1989) A parallel algorithm for matching in planar graphs, *Technical Report* TR-98-018, International

Computer Science Institute, Berkeley.

[DKK92] E. Dahlhaus, M. Karpinski and P. Kelsen (1992) An efficient parallel algorithm for computing a maximal independent set in a hypergraph of dimension 3, *Inf. Process. Lett.*, **42**, 309–313.

[DGL93] K. Diks, O. Garrido and A. Lingas (1993) Parallel algorithms for finding maximal k-dependent sets and maximal f-matchings, *Int. J. Found. Comput. Sci.*, **4** (2), 179–192.

[D52] G.A. Dirac (1952) Some theorems on abstract graphs, *Proc. London Math. Soc.*, **2**, 69–81.

[DGLL91] H. Djidjev, O. Garrido, C. Levcopoulos and A. Lingas (1991) On the maximum q-dependendent set problem, *Proc. of the International Conf. for Young Computer Scientists ICYCS91*, pp. 271–274.

[E65] J. Edmonds (1965) Paths, trees and flowers, *Can. J. Math.*, **17**, 449–467.

[E86] K. Edwards (1986) The complexity of coloring problems on dense graphs, *Theor. Comput. Sci.*, **43**, 337–343.

[E79] S. Even (1979) Graph algorithms, *Pitman.*

[FGHP93] T. Fisher, A. Goldberg, D.J. Haglin and S. Plotkin (1993) Approximating matchings in parallel, *Inf. Process. Lett.*, **46**, 115–118.

[F87] A.M. Frieze (1987) Parallel algorithms for finding hamiltonian cycles in random graphs, *Inf. Process. Lett.*, **25**, 111–117.

[F97] U. Feige (1997) Approximating the bandwidth via volume respecting embeddings, Manuscript, Weizmann Institute, Rehovot.

[FKN69] M. Fuji, T. Kasami and K. Ninomiya (1969) Optimal sequencing of two equivalent processors, *SIAM J. Appl. Math.*, **17**, 784–789.

[GT88] H. Gabov and R. Tarjan (1988) Almost optimal speed ups of algorithms for matching and related problems, in *20th Annual ACM Symp. on Theory of Computing*, pp. 514–527.

[GP88] Z. Galil and V. Pan (1988) Improved processor bounds for combinatorial problems in RNC, *Combinatorica*, **8**, 189–200.

[GP89] Z. Galil and V. Pan (1989) Parallel evaluation of the determinant and the inverse of a matrix, *Inf. Process. Lett.*, **30**, 41–45.

[GGJK78] M.R. Garey, R.L. Graham, D.S. Johnson and D.E. Knuth (1978) Complexity results for bandwidth minimization, *SIAM J. Appl. Math.*, **34**, 477–495.

[GJ79] M. Garey and D. Johnson (1979) *Computers and Intractability: A Guide to the Theory of NP-Completeness*, Freeman, San Francisco.

[GJLR96] O. Garrido, S. Jarominek, A. Lingas and W. Rytter (1996) A simple randomised parallel algorithm for maximal f-matchings, *Inf. Process. Lett.*, **57**, 83–87.

[GM87] H. Gazit and G. Miller (1987) A parallel algorithm for finding a separator in planar graphs, *28th FOCS*.

[Gi85] A. Gibbons (1985) *Algorithmic graph theory*, Cambridge University Press.

[GR88] A. Gibbons and W. Rytter (1988) *Efficient parallel algorithms*, Cambridge University Press.

[GP87] A.V. Goldberg and S.A. Plotkin (1987) Parallel $(\Delta + 1)$-coloring of constant-degree graphs, *Inf. Process. Lett.*, **25**, 241–245.

[GPST92] A. Goldberg, S. Plotkin, D. Shmoys and E. Tardos (1992) Using interior point methods for fast parallel algorithms for bipartite matching and related problems, *SIAM J. Comput.*, **21**, 148–150.

[GPV93] A. Goldberg, S. Plotkin and M. Vaidya (1993) Sublinear time parallel algorithms for matching and related problems, *J. Algorithms*, **14**, 180–213.

[GS87] M. Goldberg and T. Spencer (1987) A new parallel algorithm for the maximal independent set problem, *Proc. 28th IEEE FOCS*, pp. 161–165.

[GS89] M. Goldberg and T. Spencer (1989) Constructing a maximal independent set in parallel, *SIAM J. Discrete Math.*, **2**, 322–328.

[G82] L. Goldschlager (1982) Synchronous parallel computation, *J. ACM*, **29**, 1073–1086.

[GL95] J. Gonzales and O. Landaeta (1995) A competitive strong spanning tree algorithm for the maximum bipartite matching problem, *SIAM J. Discrete Math.*, **8** (2), 186–195.

[GK87] D. Grigoriev and M. Karpinski (1987) The matching problem for bipartite graphs with polynomially bounded permanents is in *NC*, *Proc. 28th IEEE FOCS*, pp. 166–172.

[GK90] D. Grigoriev, M. Karpinski and M. Singer (1990) Fast parallel algorithms for multivariate polynomial interpolation over finite fields, *SIAM J. Comput.*, **19**, 1059–1063.

[G56] W. Gröbner (1956) Matrizenrechnung, Oldenbourg Verl.

[GI89] D. Gusfield and R. Irving (1989) *The stable marriage problem: structure and algorithms*, MIT Press, Cambridge, MA.

[HR89] T. Hagerup and C. Rüb (1989/90) A guided tour of Chernoff bounds, *Inf. Process. Lett.*, **33**, 305–308.

[H] D.J. Haglin, Bipartite expander matching is in *NC*, to appear in *Parallel Process. Lett.*

[H56] M. Hall (1956) An algorithm for distinct representatives, *Am. Math. Mon.*, **63**, 716–717.

[H88] P. Hajnal (1988) Fast parallel algorithm for finding a Hamiltonian cycle in dense graphs, *Technical Report* 88-003-CS, Univ. of Chicago.

[HP73] F. Harary and M. Palmer (1973) *Graphical enumeration*, Academic Press.

[HM87] D. Hembold and E. Mayr (1987) Two-processor scheduling is in *NC*, *SIAM J. Comput.*, **16**, pp. 747–759.

[H96] D.S. Hochbaum, ed. (1996) *Approximation Algorithms for NP-Hard Problems*, PWS Publishing Company.

[HK73] J. Hopcroft and R.M. Karp (1973) An $O(n^{5/2})$ algorithm for maximum matching in bipartite graphs, *SIAM J. Comput.*, **2**, 225–231.

[II86] A. Israeli and A. Itai (1986) A fast and simple randomized parallel algorithm for maximal matching, *Inf. Process. Lett.*, **22**, 77–80.

[IS86] A. Israeli and Y. Shiloach (1986) An improved parallel algorithm for maximal matching, *Inf. Process. Lett.*, **22**, 57–60.

[J92] J. JáJá (1992) *An introduction to parallel algorithms*, Addison-Wesley, Reading, Massachusetts.

[JS89] M. Jerrum and A. Sinclair (1989) Approximating the permanent, *SIAM J. Comput.*, **18**, 1149–1178.

[J87] D.B. Johnson (1987) Parallel algorithms for minimum cuts and maximum flows in planar networks, *J. ACM*, **34**, 950–967.

[KKL90] M. Kao, M. Karpinski and A. Lingas (1990) Nearly optimal parallel algorithm for maximum matching in planar graphs, *Research Report* 8549-CS, University of Bonn.

[KK89] M. Kao and P. Klein (1989) Towards overcoming transitive closure bottleneck: efficient parallel algorithms for planar digraphs, *Manuscript*.

[K86] H. Karloff (1986) A Las Vegas *RNC* algorithm for maximum matching, *Combinatorica*, **6**, 387–392.

[KUW86] R.M. Karp, E. Upfal and A. Wigderson (1986) Constructing a perfect matching is in random *NC*, *Combinatorica*, **6**, 35–48.

[KW85] R.M. Karp and A. Wigderson (1985) A fast parallel algorithm for the maximal independent set problem, *J. ACM*, **32** (4), 762–773.

[K85] M. Karpinski (1985) Boolean circiut complexity of algebraic interpolation problems, Proc. CSL'88, *Lecture Notes in Computer Science*, **35**, Springer, 138–147.

[K97] M. Karpinski (1997) Polynomial time approximation schemes for some dense instances of NP-hard optimization problems, *Proc. 1st Symp. on Randomization and Approximation Techniques in Computer Science, RANDOM '97*, *Lecture Notes in Computer Science*, Springer.

[KL89] M. Karpinski and A. Lingas (1989) Subtree isomorphisms and bipartite matchings are mutually *NC*-reducible, *Inf. Process. Lett.*, **30**, 27–32.

[KW88] M. Karpinski and K. Wagner (1988) The computational complexity of succinct multi-vertex representation of graphs, *Z. Oper. Res.*, **32**, 201–211.

[KW97a] M. Karpinski and J. Wirtgen (1997) NP-hardness of the bandwidth problem on dense graphs, *Research Report* No. 85176-CS, University of Bonn.

[KW97b] M. Karpinski and J. Wirtgen (1997) On approximation hardness of the bandwidth problem, *Technical Report* No. TR97-041, ECCC.

[KWZ97] M. Karpinski, J. Wirtgen and A. Zelikovsky (1997) An approximation algorithm for the bandwidth problem on dense graphs, *Research Report* No. 85164-CS, University of Bonn.

[KZ97] M. Karpinski and A. Zelikowski (1997) New approximation algorithms for the Steiner tree problem, *J. Comb. Opt.*, **1**, 47–65.

[K63] P.W. Kasteleyn (1963) Dimer statistics and phase transitions, *J. Math. Phys.*, **4**, pp. 287-293.

[K67] P.W. Kasteleyn (1967) Graph Theory and Crystal Physics, in *Graph Theory and Theoretical Physics*, ed. F. Havary, Academic Press, New York, pp. 43–110.

[Ke92] P. Kelsen (1992) On the parallel complexity of computing a maximal independent set in a hypergraph, *24th ACM Symp. on the Theory of Computing*, pp. 339–350.

[K94] P. Kelsen (1994) An optimal parallel algorithm for maximal matchings, *Inf. Process. Lett.*, **52**, 223–228.

[KRS93] C. Kenyon, D. Randall and A. Sinclair (1993) Matchings in lattice graphs, *STOC'93*, pp. 728–737.

[K93] P. Klein (1993) Parallel algorithms for chordal graphs, in *Synthesis of Parallel Algorithms*, San Mateo, California.

[KR87] P. Klein and J. Reif (1988) An efficient parallel algorithm for planarity, *J. Comput. Syst. Sci.*, **37**, 190–246.

[Ko92] D. Kozen (1992) *The design and analysis of algorithms*, Springer Verlag.

[KVV85] D. Kozen, U. Vazirani and V. Vazirani (1985) *NC* algorithms for comparability graphs, interval graphs and testing for unique perfect matching, *Proc. FST&TCS, LNCS 206*, pp. 496–503.

[LPV81] G. Lev, M. Pippenger and L. Valiant (1981) A fast parallel algorithm for routing in permutation networks, *IEEE Trans. Comput.*, **C-30**, 93–100.

[LB] Y.D. Liang and N. Blum, Circular convex bipartite graphs: maximum matching and Hamiltonian circuits, to appear in *Inf. Process. Lett.*

[LP81] W. Lipski and F. Preparata (1981) Efficient algorithms for finding maximum matching in bipartite graphs and related problems, *Acta Inf.*, **15**, 329–346.

[LT79] R. Lipton and R. Tarjan (1979) Applications of planar separator theorem, *SIAM J. Appl. Math.*, **36**, 177–189.

[L74] C.H.C. Little (1974) An extension of Kasteleyn's method of enumerating the 1-factors of planar graphs, in *Combinatorial Mathematics, Proc. 2nd Australian Conference*, ed. D. Holton, *Lecture Notes in Mathematics*, **403**, Springer Verlag, pp. 63–72.

[L75] L. Lovász (1975) Three short proofs in graph theory, *J. Comb. Theor.*, B, **19**, 111–113.

[L79a] L. Lovász (1979) Determinants, matchings, and random algorithms, *Fundamentals of Computation Theory, FCT '79*, ed. L. Budach, Akademie-Verlag, Berlin, pp. 565–574.

[L79b] L. Lovász (1979) *Combinatorial Problems and Exercises*, North-Holland.

[LP86] L. Lovász and M. Plummer (1986) Matching Theory, *Mathematical Studies, Annals of Discrete Mathematics*, Vol. 25, North-Holland, Amsterdam.

[L86] M. Luby (1986) A simple parallel algorithm for the maximal independent set problem, *SIAM J. Comput.*, **15**, 1036-1053.

[MV97] M. Mahajan and V. Vinay (1997) A combinatorial algorithm for the determinant, *SODA97*, pp. 730–738.

[MS92] E. Mayr and A. Subramanian (1992) The complexity of circuit value problem and network stability, *J. Comput. Syst. Sci.*, **44**, 302–323.

[MV80] S. Micali and V. Vazirani (1980) An $O(\sqrt{|V|}\ |E|)$ algorithm for finding maximum matching in general graphs, *Proc. 21st IEEE FOCS*, pp. 17–27.

[MN95] G. Miller, J. Naor (1995) Flow in planar graphs with multiple sources and sinks, *SIAM J. Comput.*, **24**, 1002–1017.

[M78] H. Minc (1978) *Permanents*, Addison Wesley.

[MR95] R. Motwani and P. Raghavan (1995) *Randomized Algorithms*, Cambridge University Press.

[Mu05] T. Muir (1905) *The Theory of Determinants*, Dover, New York.

[MVV87] K. Mulmuley, U. Vazirani and V. Vazirani (1987) Matching is as easy as matrix inversion, *Combinatorica*, **7** (1), 105–131.

[N93] S. Naor (1993) Probabilistic methods in computer science, *Notes.*

[OA89] C.N.K. Osiakwan and S.G. Akl (1989) Optimal parallel algorithms for b-mathings in trees, *Proc. Int. Symp. on Optimal Algorithms, Lecture Notes in Computer Science*, **401**, 274-308.

[P92] V.Th. Paschos (1992) A $(\Delta/2)$-approximation algorithm for the maximum independent set problem, *Inf. Process. Lett.*, **44**, 11–13.

[P85] V. Pan (1985) Fast and efficient algorithms for exact inversion of integer matrices, *Lecture Notes in Computer Science*, **206**, 504–521.

[PS82] C.H. Papadimitriou and K. Steiglitz (1982) *Combinatorial optimization: algorithms and complexity*, Prentice-Hall, NJ, Englewood Cliffs.

[PU86] D. Peleg and E. Upfal (1986) The token distribution problem, *Proc. 27th IEEE FOCS*, pp. 418–427.

[PU87] D. Peleg and E. Upfal (1987) Constructing disjoint paths on expander graphs, *Proc. 19th ACM STOC*, pp. 264–273.

[PS78] F.P. Preparata and D.V. Sarwate (1978) An improved parallel processor bound in fast matrix inversion, *Inf. Process. Lett.*, **7**, 148–150.

[PS97] H.J. Prömel and A. Steger (1997) *RNC*-approximation algorithms for the Steiner problem, *Proc. STACS'97.*

[QD84] M.J. Quinn and N. Deo (1984) Parallel graph algorithms, *ACM Comput. Surv.*, **16**, 3.

[RV89] M. Rabin and V. Vazirani (1989) Maximum matching in general graphs through randomization, *J. Algorithms*, **10**, 557–567.

[SM86] B. Schieber and S. Moran (1986) Slowing down sequential algorithms for obtaining fast distributed and parallel algorithms: maximum matching, *Proc. 15th Annual Symp. on Principles of Distributed Computing*, pp. 282–292.

[S80] J.T. Schwartz (1980) Fast probabilistic algorithms for verification of polynomial identities, *J. ACM*, **27**, 701–717.

[SW96] R. Sharan and A. Wigderson (1996) A new approach to the perfect matching problem, *Proc. Israel Symp. on Theoretical Computer Science.*

[SV82] Y. Shiloach and U. Vishkin (1982) An $O(n^2 \log n)$ parallel max-flow Algorithm, *J. Algorithms*, **3**, 128–146.

[SM95] T. Shoudai and S. Miyano (1995) Using maximal independent sets to solve problems in parallel, *Theor. Comput. Sci.*, **148**, 57–65.

[Si93] A. Sinclair (1993) *Algorithms for random generation and counting: a Markov chain approach, Monograph Series Progress in Theoretical Computer Science*, Birkhauser, Boston.

[S86] D. Soroker (1986) Fast parallel algorithms for finding hamiltonian paths and cycles in a tournament, *Report* No. UCB/CSD 87/309, University of California, Berkeley.

[Sp93] P. Spirakis (1993) PRAM models and techniques: Part II, in *Lectures on Parallel computations*, ed. A. Gibbons and P. Spirakis, Cambridge University Press, pp. 41–66.

[T47] W.T. Tutte (1947) The factorization of linear graphs, *J. London*

Math. Soc., **22**, 107–111.

[V89] V. Vazirani (1989) *NC* algorithms for computing the number of perfect matchings in $K_{3,3}$-free graphs and related problems, *Inf. Comput.*, **80**, 152–164.

[V91] V. Vazirani (1991) Parallel graph matching, chapter 18 in *Synthesis of parallel algorithms*, ed. J. Reif, Kaufman-Morgan.

[W91] J. Wein (1991) Las Vegas *RNC* algorithms for unary weighted perfect matching and T-join problems, *Inf. Process. Lett.*, **40**, 161–167.

[W93] D.J.A. Welsh (1993) *Complexity: Knots, Colourings and Counting, Lecture Note Series*, **186**, London Mathematical Society, Cambridge, pp. 1–163.

[Z79] R.E. Zippel (1979) Probabilistic algorithms for sparse polynomials, in *EUROSAM 79, Lecture Notes in Computer Science*, **72**, pp. 216–226.

Index